Preface

Photosynthesis—the capture of light energy by living organisms—is a simple enough concept, but its investigation draws on the resources of disciplines from all fields of science. The aim of this text is to provide a clear, stimulating and essentially affordable coverage for undergraduate students of biology.

The activity of science is debate and practical experiment; its product is a body of propositions which at any given time reflects the judgment and prejudices of those taking part. The value of a proposition is related to the conceivable alternatives, and writing it down without its context creates the false impression that science progresses by compilation of an increasing list of absolute truths. It does not; the facts and figures presented in the following pages have no intrinsic value unless they can be used by the reader to support an argument or point of view. In short, the reader is urged to respond 'So what?' to every item. Secondly, ideas—like other foods—should be date-stamped; science is inseparable from its history. I have set out time-charts to represent the evolution of our understanding in certain areas.

I have assumed that the reader is pursuing a course with a content of biochemistry, microbiology and plant science, or has access to basic texts. I have assumed also that common methods such as spectrophotometry, chromatography and electrophoresis, as well as the techniques of molecular biology, will be either part of the same course or in active use nearby.

If arranging the material on the basis of time-scales (and I acknowledge the prior use of this approach by R.K. Clayton and others) I have biased the text towards biophysics and biochemistry. I hope the biologists will permit this approach; it is both concentric and centrifugal and begins with matters which are simple, at least in the sense of being small-scale. For example, at the large-scale end, geographers may note that estimates of global ocean productivity, derived from satellite-based observations of fluorescence, depend on the molecular-scale connectivity of antenna phycobilins in cyanobacteria, which may not be constant.

The references to the literature have been chosen largely to provide accounts of crucial experiments and more esoteric techniques. Several multi-volume series have recently appeared in biochemical or plant science subjects, that have devoted volumes to photosynthesis, and the periodical *Trends in Biochemical Science* cannot be too highly praised for its succinct, topical and accessible articles in this field. Regular reviews appear in, for example, *Annual reviews in Plant Physiology*, whilst for the intending specialist, the proceedings of the triennial International Congresses on Photosynthesis are invaluable in providing a flavour of the current spread of research activity.

I am indebted to my colleagues who have helped and advised during the preparation of the text. I thank Professor W. Wehrmeyer of the University of Marburg for his electron micrographs, Miss Angela Jones and Miss Denise Smith of the Department of Medical Illustration and Mr W.T. Hewitt of the Computer Graphics Unit, Manchester University, for their invaluable assistance.

M.W.

During the preparation of this text, I was delighted to hear of the award of the 1988 Nobel Prize for Chemistry to J. Deisenhofer, R. Huber and H. Michel. Their X-ray crystallographic resolution of the structure of the reaction centre from the purple bacterium *Rhodopseudomonas viridis* was a great stimulus to all who work in this field; I expect the pace of discovery to continue in the last decade of this millennium.

Contents

To Kathleen

CHAPTER ONE

PHOTOSYNTHESIS
The capture of light energy by living organisms

Life needs a supply of energy. Living things maintain an elaborate structure and organisation, they repair damage and grow, they respond to stimuli and reproduce their kind. They do chemical, osmotic, electrical and mechanical work and may give out heat, sound and light. Where does the energy come from? There are only two sources available: chemical energy in the environment, and light. Chemical energy occurs in the form of reactable chemical substances; we ourselves as animals look for food (such as glucose) which we cause to react (by means of a series of steps) with oxygen in the air. The other energy source is light. Photosynthesis is the lifestyle whereby a plant or bacterium captures light by absorption in a pigment, and converts the energy to its own use.

The importance of photosynthesis can hardly be overstated. All the oxygen of the atmosphere has been and continues to be provided by photosynthesis in green plants, and plants are by far the most important starting points for the food chains on which animals depend. Light energy in plants becomes chemical energy for animals.

The only sources of chemical energy that are not produced by other living organisms are the chemically-reducing, sulphurous gases that emerge from volcanic vents, and the nitrates (oxidising) originating from lightning discharges. These materials, today, make an infinitesimal contribution to the balance of the earth's energy, although in the distant past electrical discharges in a reducing atmosphere may have provided the organic mixture from which life began.

The biosphere is closely balanced. Oxygen is produced by photosynthesis in green plants at a rate of 5.1×10^{14} kg per year. The atmosphere contains 1.1×10^{18} kg of oxygen, and remains constant: consumption is equal to production. If some factor such as industrial pollution were to severely diminish the rate of photosynthesis, the level of oxygen would fall critically in a few hundred years. Even more serious is the balance of carbon dioxide,

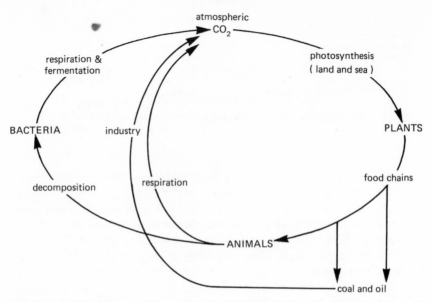

Figure 1.1. The carbon cycle. The cycle shows the biological conversion of atmospheric and oceanic carbon dioxide into plant biomass, then into animal biomass and carbonaceous fossil deposits, and back into the atmosphere by combustion and respiration. The speed with which the carbon dioxide level in the atmosphere changes is a matter of current concern.

because there are only 5×10^{16} kg (mostly dissolved in the sea), consumed by photosynthesis mainly in green plants at an annual rate of 7×10^{14} kg (Riley, 1944). This is the well-known carbon cycle (Figure 1.1). The increase in the level of carbon dioxide due to industrial burning of fossil fuels is expected to cause severe environmental problems, since the 'greenhouse' effect will lead to higher average climatic temperatures. That is an optimistic forecast, assuming that higher rates of photosynthesis will largely counteract the increase. Loss of photosynthetic capacity would allow a much sharper rise in the atmospheric carbon dioxide content.

1.1 Photosynthesis in historical context

A time-chart of discoveries and ideas is presented in Table 1.1, from which some names stand out. Joseph Priestley observed in 1771 that green plants differed from animals in their effect on the air in a closed space. Animals such as mice altered the air in the same way as did a burning candle, and the air was soon unable to support further the life of mice or the burning of

Table 1.1 Time chart of early photosynthesis research.

Date	Investigators	Observation
17C	J.B. van Helmont	Mass gain of tree exceeds mass loss of soil: mass gain comes from water
1727	Stephen Hales	Biological gas exchanges Air is chief plant nutrient
Mid 18C	J. Priestley J. Ingen-Housz	Plants in confined space regenerate air—compared with animals and candles —only green parts, in light
Late 18C	Jean Senebier	Identified carbon dioxide as chief plant nutrient
	A. Lavoisier	Gravimetric study of combustion: oxygen theory connects combustion, respn. and photosynth.
1804	N.T. de Saussure	Gravimetry of photosynthesis CO_2 *and* water required: equation
1817	Pelletier and Caventou	Isolated and named chlorophyll
1837	H. von Mohl	Described the chloroplast
1845	Robert Meyer	Interpreted photosynthesis as capture of light energy
1864	J. Sachs	Observed starch grains formed by leaves in light
1880	Engelmann	Motile bacteria congregate around filamentous algal cell adjacent to chloroplast; illuminated by a spectrum, red and blue were preferred to green: light and Chl produce O_2
1883	A.F.W. Schimper	(Chloro)plastids divide: symbiotic in plant cells
1883	A. Meyer	Described chloroplast grana
1905	F.F. Blackman	Observed light-saturation curve 'Light' and 'dark reactions'
1932	R. Emerson and W. Arnold	Saturating flashes produce one molecule O_2/2500 Chl 2500 Chl = photosynthetic unit
1939	R. Hill	Isolated Chlp + Fe^{3+} produce O_2 Hill reaction in chloroplasts
1954	A. Benson, M. Calvin	Products of photosynthesis in *Chlorella* cells with $^{14}CO_2$ Reductive pentose cycle
1954	D.I. Arnon	CO_2 uptake supported O_2 production in isolated Chlp: Chlp self-sufficient
1954	D.I. Arnon	Spinach Chlp produce ATP 'photophosphorylation'
1954	A.W. Frenkel	*R. rubrum* chromatophores produce ATP—crucial for photosynthesis

The table sets out the milestones in the early history of photosynthesis. Sources: Rabinowitch, E.I. (1945), Arnon, D.I. (1977) and Hoober, J.K. (1984).

candles. On the other hand a green plant, illuminated in the altered air, soon restored it to its original good condition so that once again mice could survive and candles burn in it. At that time the chemical theory of combustion was based on a hypothetical substance called phlogiston; in a fire, phlogiston was supposed to leave the burning materials and fill up the air. Priestley was unable to make much further progress because the 'phlogiston theory' of combustion was inadequate, and Jan Ingen-Housz refined the observations and showed that mice, and candles, consumed oxygen and produced carbon dioxide, while illuminated green plants consumed carbon dioxide and produced oxygen. The activity of the animal was known as respiration, and identified as a combustion reaction by A. Lavoisier in 1780. It was later summarised by the equation:

$$C_6H_{12}O_6 + 6O_2 \rightarrow 6CO_2 + 6H_2O.$$

N.T. de Saussure showed in 1804 that the fixation of carbon dioxide by green plants could be described by an analogous equation:

$$6CO_2 + 6H_2O \xrightarrow{\text{light}} 6O_2 + C_6H_{12}O_6.$$

This equation is now known as the equation of photosynthesis, although the word 'photosynthesis' did not appear until 1893–8. Up to that time the word 'assimilation' had been used, by analogy with animal digestion. The requirement for light energy, and the fixation of carbon dioxide, were regarded as inseparably part of the same process.

The equation of respiration is effectively a description of combustion, and the energy released by the combustion of the carbohydrate accounts for the warmth and the animal's activity (work done). The fact that the photosynthesis equation appears as a reversal of respiration, requiring an input of energy, and that light (a form of energy) is needed, provided the obvious conclusion that the green plant was able to harness light energy to drive an energy-consuming process.

The absorption and conservation of light energy, and the fixation of carbon dioxide, were separated when R. Hill, in 1939, showed that isolated (but in fact damaged) chloroplasts could produce oxygen at the expense of an artificial oxidant (a ferric-iron complex). Ten years later M. Calvin's group elucidated the enzyme-catalysed reactions whereby carbon dioxide is fixed, and ten years later still undamaged chloroplasts were isolated which were competent in both activities at natural rates.

1.2 Photosynthesis in a comparative context

Photosynthesis as first defined applied only to the light-dependent gas exchange carried out by green plants. The subject widened when C.B. van Niel reported in 1930 the discovery that the anaerobic groups of green and purple bacteria needed light for their growth. Some groups could fix carbon dioxide without any concomitant production of oxygen; in these cases instead of water a different environmental reducing agent such as hydrogen sulphide (H_2S) was oxidised to (for example) elemental sulphur giving a different form of the equation for photosynthesis:

$$6CO_2 + 12H_2S \xrightarrow{\text{light}} 6S_2 + C_6H_{12}O_6 + 6H_2O.$$

Other purple (and later green) bacteria were found to use light, to drive the fixation not of carbon dioxide, but of organic nutrients such as acetate or malate. The overall chemical process still required energy, which accounted for the need for light, but was no longer necessarily a reduction process. If green-plant photosynthesis is in a sense a reversal of respiration, then the process carried out by these bacteria can be regarded as a light-driven reversal of fermentation. These light-dependent bacteria contain, not chlorophyll as found in green plants, but a close derivative, bacterio-chlorophyll.

'Photosynthesis' now applies to any habit in which the energy for the life of an organism is obtained from light.

'Lithotrophic bacteria' were discovered by S. Winogradsky in 1880–90. This group was later shown to incorporate carbon dioxide by the same process as green-plant photosynthesis, except that light was not required (the bacteria are colourless) and the necessary energy was obtained from a concomitant oxidation of an inorganic material such as ferrous ions (Fe^{2+}), hydrogen sulphide, nitrite (NO^{2-}) or nitrate (NO^{3-}). This habit is known as 'chemosynthesis' by analogy with photosynthesis according to the equations:

$$6CO_2 + 12H_2S \xrightarrow[\text{Thiobacillus thioxydans}]{H_2S + O_2 \rightarrow H_2O + S} C_6H_{12}O_6 + 6S_2 + 6H_2O$$

and

$$6CO_2 + 24Fe^{2+} + 18H_2O \xrightarrow[\text{T. ferrooxydans}]{Fe^{2+} + O_2 \rightarrow Fe^{3+}} C_6H_{12}O_6 + 24Fe^{3+} + 24OH^-.$$

The inorganic reductant provides the hydrogen incorporated into the carbohydrate, and the oxidation by oxygen provides the energy for the carbon-dioxide fixation reaction. The lithotrophic bacteria fix carbon dioxide by a metabolic mechanism almost identical to that of the green plant, but they consume oxygen instead of requiring light energy. Carbon dioxide fixation in general is not necessarily photosynthetic, and photosynthesis does not necessarily include the fixation (assimilation) of carbon dioxide.

Several terms are used to cover these various lifestyles:

Phototrophy indicates that the energy source is light. This covers green plants and the green and purple bacteria;

Chemotrophy indicates that the organism carries out chemical reactions to obtain energy (from its environment). This describes animals and most bacteria including the lithotrophs;

Autotrophy describes the situation where the sole source of carbon for the growth of the organism is carbon dioxide, as in green plants, some green and purple bacteria, and lithotrophic bacteria;

Heterotrophy (or organotrophy) describes the use of organic materials such as carbohydrate, acetate and malate, as the source of carbon, as in animals, and most bacteria including some green and purple types;

These four terms overlap, and can be combined. Thus photoheterotrophy describes photosynthesis (by bacteria) in which organic carbon is used, and photoautotrophy is photosynthesis using carbon dioxide, as in green plants. Chemoautotrophy is synonymous with lithotrophy (only); it is possible that an organism could obtain energy from oxidation of an organic substance while relying on carbon dioxide for the carbon supply, but no examples are known. A green photosynthetic bacterium that appeared to use organic material solely as the reductant for carbon dioxide was in fact shown to be a mixed culture. Chemoheterotrophy describes animals, plants in darkness and microorganisms performing respiration or fermentation. The above terms are best kept as labels for habits, not organisms, since many bacteria are adaptable. Some photosynthetic purple bacteria can live chemoautotrophically (lithotrophically), photoheterotrophically or photoautotrophically. Species that are apparently closely related may differ in their range of lifestyles.

Photosynthesis is now synonymous with phototrophy. There is, however, a distinction to be drawn between two forms of photosynthesis. All the groups of plants and photosynthetic bacteria referred to above are clearly related; light is captured by some form of chlorophyll and drives electron transport through a series of steps, some of which are homologous with those of respiration in the mitochondria of eukaryotic cells, and the cell membranes of aerobic bacteria. It is important that the production of adenosine triphosphate (ATP, the 'energy currency of the cell') is produced in much the same way in mitochondria as in chloroplasts and bacteria, that is, by means of a chain of electron-transport carriers that store their energy in hydrogen-ion transport.

However, W. Stoeckenius, in 1960, described a novel bacterium, *Halobacterium halobium*, which was able to use light energy whenever its normal respiratory (chemoheterotrophic) activity was restricted by a storage of food or oxygen. It was shown that no chlorophyll derivatives *Shortage?* were present and the pigment used for absorbing light was retinal (a carotene derivative), carried in the protein bacteriorhodopsin. No electron transport was involved, but H^+ transport (pumping) was achieved directly by bacteriorhodopsin itself. The mechanism of ATP production, however, corresponded closely with that of mitochondria, chloroplasts, and the other bacteria described above. We may usefully distinguish 'chlorophyll-based photosynthesis' from 'retinal-based photosynthesis'. The distinction is valid since the latter is only found in the archaebacteria, and the former in the eubacteria; the two groups have relatively little in common.

Studies covering all the ATP-producing systems (both types of photosynthesis, and respiration, in bacteria, chloroplasts and mitochondria) are included in the term 'bioenergetics', which also covers the utilisation of ATP by, for example, contractile proteins and ion-pumps, and the bioenergetic approach to the study of photosynthesis has proved most valuable (see Table 1.2).

Probably the most simple lifestyle is that of fermentation. It depends on the organism having a foodstuff that can undergo ATP-producing reactions without a need for oxygen. Such organisms are, for example, obligate anaerobic bacteria such as *Clostridium* (which do not possess an electron transport or cytochrome chain) and facultative anaerobes such as yeast. In the case of yeast the ethanolic fermentative reaction of glucose can be summarised:

$$C_6H_{12}O_6 \rightarrow 2CO_2 + 2C_2H_5OH.$$

One of the functions of ATP is to enable reactions to proceed which build up more complex materials, that is, to support the growth of the organism. Fermentative breakdown of part of the foodstuffs allows the incorporation of the other part. Anaerobic bacteria may generate hydrogen gas or carbon dioxide in order to achieve the required oxidation-reduction balance of the products.

Table 1.2 The bioenergetic approach to biology. Comparative features of ATP production in various organisms.

Example	Habit	ATP synthase particle	Electron-transport chain	Source of ATP	Source of NAD(P)H
Clostridium	Fermenter	$-$*	$-$**	S-level	Substrate
Strep. faecalis	Fermenter	$-$*	$-$	S-level	Substrate
H. halobium					
(1) Resp. mem.	Respiration	$+$	$+$	H-ET	Substrate
(2) Purple mem.	Photosynthesis	$+$	$-$	H-direct	None
R. rubrum					
(1) Anaerobic	Photosynthesis	$+$	$+$	H-ET	Inorganic e.g. H_2
(2) Aerobic	Respiration	$+$	$+$	H-ET	Substrate
Cyanobacterium					
and chloroplast	Photosynthesis	$+$	$+$	H-ET	Water
E. coli					
and mitochondrion	Respiration	$+$	$+$	H-ET	Substrate

'Electron transport chain' implies a cytochrome *bc* (quinol dehydrogenase) complex. Under 'source of ATP' S-level denotes 'substrate level', that is, ATP produced by a metabolic enzyme; H denotes F-ATPase coupled to H^+ ions pumped by electron transport (ET) or by bacteriorhodopsin (direct).
*These fermenters possess a F-ATPase particle which is used to create a proton gradient at the expense of ATP. ATP can be formed by experimentally increasing the proton gradient.
**Electron transport and ATP synthase action is not ruled out in the case of the facultative autotroph *C. thermoaceticum*.

Fermentation is not needed when ATP is supplied by the H^+-dependent F-ATPase. The circulation of H^+ ions may be driven chemotrophically by electron transport from a reductant to an oxidant. The reductant may be organic (chemoheterotrophy), and the oxidant either oxygen, as in the case of normal respiration, or an inorganic ion such as sulphate or nitrate in the cases of some bacteria. Alternatively the reductant may be inorganic and the oxidant oxygen, as in the (chemo)lithotrophic bacteria.

It is a feature of the electron transport chain that it can to some extent function in reverse, so that given a supply of ATP, the relatively weak inorganic reductants can give rise to the more powerful reductants that allow the reduction of carbon dioxide. This provides for autotrophic growth in purple photosynthetic bacteria, and in lithotrophic bacteria. *Halobacterium halobium* is an oddity here: it can drive its F-ATPase either by normal respiration in its normal membrane, or by bacteriorhodopsin in the 'purple membrane' patches. It cannot fix carbon dioxide, both because it cannot use inorganic reductants, and because the essential enzymes are absent.

In chlorophyll-based photosynthesis the electron transport chain is driven by light, not by the energy of reaction of external oxidants and reductants. In the simplest arrangement (in the purple bacteria) the only direct product of the light-driven electron transport is ATP, and the organisms grow by incorporating organic nutrients, releasing hydrogen or carbon dioxide as necessary. However, in some cases purple bacteria (especially the 'sulphur' bacteria) can connect environmental reductants such as sulphide, thiosulphate or molecular hydrogen to the electron transport chain, and this provides (ATP-assisted) reducing power for the fixation of carbon dioxide. In some cases again (the 'non-sulphur' bacteria, some of which can paradoxically use sulphide), there is an ability to dispense with the light-utilising system, and develop oxidases connecting the electron transport chain to molecular oxygen. This allows them to respire organic nutrients like many other bacteria and animals, or to carry out chemoautotrophy. The green non-sulphur bacteria are in many respects similar to their purple counterparts, but the green sulphur bacteria differ from both in that they are unable to grow heterotrophically, but necessarily use a reductant such as sulphide to fix carbon dioxide.

The bacteriochlorophyll-based photochemical systems of the purple and green non-sulphur bacteria are closely similar, but that of the green sulphur bacteria differs markedly from them. In two other groups of prokaryotes (cells without nuclei), the Prochlorophyta and the Cyanobacteria (formerly the blue-green algae), there are two photosystems, one constructed on similar lines to that of the purple bacteria, and one resembling that of the green sulphur bacteria. They both contain chlorophyll, not bacteriochlorophyll, and both use water as an oxidant, producing molecular oxygen, a gas which is toxic to most of the green and purple sulphur bacteria.

Green plants (which belong to the group of Eukaryotes, cells with nuclei) differ from prochlorophyta and cyanobacteria in that the eukaryotes have chloroplasts in their cells; chloroplasts closely resemble prokaryote cells. Green plant cells also have mitochondria, and mitochondria in many ways resemble aerobic bacteria. Eukaryote cells therefore depend on 'prokaryotic' organelles, chloroplasts in green plants and mitochondria in plants and animals, for their energy supply.

Our study of photosynthesis necessarily covers a variety of organisms and their lifestyles. Table 1.3 provides a list of the organisms that will be mentioned, and their classification. As hinted above, strong relationships exist between the different groups. Such relations accord with the principle of evolution, although when dealing with bacteria, the possession of related

Table 1.3 Some photosynthetic organisms and their classification.

PROKARYOTES

Archaebacteria:	*Halobacterium halobium*	Light capture by bacteriorhodopsin
Eubacteria:		
Chlorobiaciae:		Green sulphur bacteria
	Chlorobium (Cb) limicola f. thiosulfatophilum	
	Prosthecochloris aestuarii	
Filamentous photosynthetic bacteria:		Green non-sulphur bacteria
	Chloroflexus aurantiacus	
Unassigned:	*Heliobacterium chlorum*	
Purple bacteria:		
Rhodospirillaceae		Former Athiorhodaceae
	Rhodospirillum (R) rubrum	
	Rhodobacter (Rb) spheroides	
	Rb. capsulatus	
	Rhodocyclus gelatinosus	
	Rhodopseudomonas (Rps) viridis	
Chromatiaceae		Former Thiorhodaceae
	Chromatium	
Cyanobacteria:		Blue-green algae
	Anacystis nidulans	
	Mastigocladus laminosus	
Prochlorophyta:	*Prochloron*	

EUKARYOTES

Algae:		
Rhodophyta:		Red algae
	Porphyridium cruentum	
Chromophyta:		Brown algae, diatoms, etc.
	Ochromonas danica	
Chlorophyta:		Green algae
	Ankistrodesmus	
	Chlorella	
	Euglena	
	Mougeotia	
	Scenedesmus	
Bryophyta:		Liverworts and mosses
	Marchantia polymorpha	
Angiospermae:		Flowering plants
	Digitaria sanguinalis	Crabgrass
	Hordeum vulgare	Barley
	Nicotiana tabacum	Tobacco
	Pisum sativum	Pea
	Phaseolus vulgaris	Bean
	Spinacaea oleracea	Spinach
	Zea mays	Maize

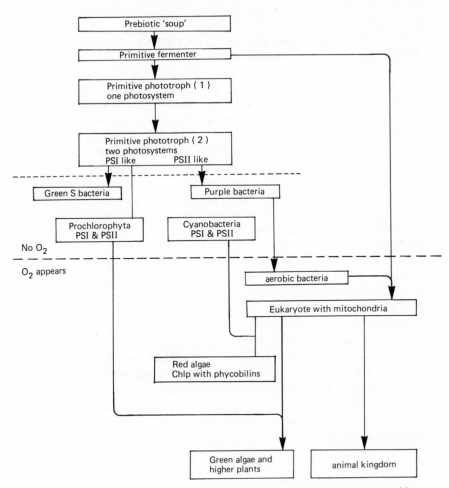

Figure 1.2. The relationships of phototrophic and chemotrophic organisms, arranged in an imaginary evolutionary pattern. The four stages above the dotted line are entirely hypothetical and represent life forms that existed about 3300 million years ago. All the other lifeforms are in existence today. The dashed line represents the transition from a reducing to an oxidising atmosphere brought about by the appearance of oxygenic photosynthesis in the cyano-bacteria, and prochlorophyta, about 2200 million years ago.

12   PHOTOSYNTHESIS

biochemical processes is a shaky foundation for establishing evolutionary relationships.

Speculating on the evolution of photosynthesis, geological structures are known (stromatolites) which are almost certainly due to photosynthetic bacteria 3500 million years ago, an epoch when the earth's atmosphere contained no oxygen and was chemically reducing. It cannot at present be determined whether there was at that time any local or general organic material capable of supporting chemoheterotrophy; there may have been a fermentative life-form that has been lost. The (chlorophyll-based) photosynthetic bacteria of the time can be readily imagined to give rise to oxygen-evolving blue-green bacteria (2500 to 3000 million years ago). As the oxygen content of the biosphere increased the respiratory apparatus may have evolved, perhaps primarily as a means of protection against the new toxic gas, and a population of non-photosynthetic, respiring chemohetero-trophic bacteria arose and diversified giving rise to, amongst other forms, the archaebacterial, retinal-based phototroph. Stretching the imagination only slightly, the respiring eubacteria could have invaded some cell (such as the hypothetical fermentative form above) giving rise to the mitochondria found in eukaryotic cells, and a second invasion, from cyanobacteria or prochlorophytes would account for the origin of chloroplasts, in something resembling the red and green algae respectively, from which evolved the present world of green plants. Such a scheme is shown in Figure 1.2. Evidence for evolutionary relationships comes from the comparison of proteins and nucleic acids. Proteins such as ferredoxin, or the cytochrome *bc* complex (see, e.g., Gabellini, 1988), are valuable because their crucial functions are not expected to change, and mutations in the protein sequence occur slowly, because few will be viable. Therefore the more closely related two organisms are, the more the structures of the proteins should agree. The same applies to the sequences of bases in the genes for ribosomal RNA in the chloroplast genome.

CHAPTER TWO

THE ORGANISATION OF PHOTOSYNTHETIC STRUCTURES

It was pointed out by Ingen-Housz that only the green parts of plants carried out the re-oxygenation of air. Greenness is due to the presence of the pigment chlorophyll, which is only found in the internal membranes of chloroplasts. In this section we shall see that membranes and their structure are central to all forms of photosynthesis.

2.1 Biological membranes

A living cell is composed of protoplasm which is principally a solution of proteins (the cytosol), in a total concentration of some 20%, with particles, fibrils and (in eukaryotes) membrane-bounded organelles contained in it. The cell is itself bounded by a membrane known as the plasmalemma or cell membrane. Other membrane structures are found in the cells of animals and green plants. All the membranes of the cell are to be regarded as fluid bilayer structures composed of polar lipids. The hydrophobic hydrocarbon 'tails' of the lipid molecules are directed inwards to the centre of the membrane, and the polar 'head groups' (commonly phosphate or carbo-hydrate) are on the outside in contact with the aqueous phases (Figure 2.1). The composition is constant for a given membrane, but there is consider-able variety between different membranes even in the same cell.

The lipid bilayer membrane provides a barrier to the diffusion of hydrophilic solutes, such as ions or metabolic intermediates, from one aqueous phase to the other. It is also an electrical insulator of considerable dielectric strength. It possesses virtually no mechanical strength (being a fluid); since the cell usually has a higher osmotic pressure than the external medium, the plasmalemma is often supported by a cell wall.

Most of the distinctive properties of individual membranes are conferred by protein molecules embedded in the bilayer (intrinsic membrane proteins)

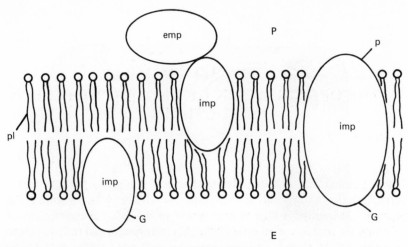

Figure 2.1. The 'fluid mosaic' model of biological membrane structure, proposed originally by Singer and Nicolson (1972). Polar lipids (pl) (phospholipid or galactolipid) are arranged in a bilayer, tails to tails, and intrinsic membrane proteins (imp) may be located in either leaflet, or traverse the entire membrane. Extrinsic membrane proteins (emp) are attached to an intrinsic anchorage. Glycosylation (G) and phosphorylation (P) are found on the E- and P-sides respectively (see Figure 2.2).

or attached to the surface (extrinsic membrane proteins). The external surface particularly of plasmalemma membranes carries carbohydrate and glycolipid. At least some of these extrinsic materials are attached to intrinsic proteins. Some of the proteins are embedded in such a way that they are accessible to the aqueous phases on both sides of the membrane (membrane-spanning proteins), while others are located in either the outside or the inside leaflet. This is the 'fluid mosaic' concept of membrane structure developed by Singer and Nicolson (1972). The total effect is that the membrane is asymmetric, the outside being different in its accessible groups from the inside surface. Intrinsic proteins always have the same orientation; they can often rotate about an axis perpendicular to the membrane, but cannot tumble. When comparing different membranes, it is observed that in many cases the membrane has one surface in contact with protoplasm and the other surface in contact with a solution much less rich in protein. The protoplasmic surface has been labelled by electron-microscopists the P-surface, and the other surface which in the plasma-lemma is in contact with the external medium is the E-surface (Branton *et al.*, 1975). Virtually all biological membranes can be regarded as separating

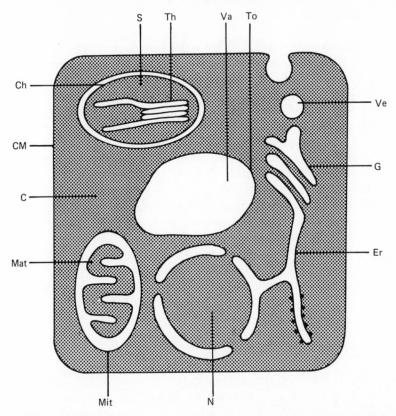

Figure 2.2. Diagram of cellular organisation, to show that all cell spaces may be regarded as 'E' (related to the external medium; 'exoplasmic space' of C. de Duve) or 'P' (protoplasmic). The P spaces are shaded. Symbols: Ch, chloroplast (envelope); CM, cell membrane (plasmalemma); Cr, crista; Er, endoplasmic reticulum; G, Golgi apparatus; Mit, mitochondrion (envelope); Mat, matrix; N, nucleus; S, stroma; Th, thylakoids; To, tonoplast membrane; Va, vacuole and Ve, endo- or exocytotic vesicles.

E- and P-phases, although it may seem strange to regard the spaces inside the chloroplast, or between the nuclear membranes, as in any way 'external' (see Figure 2.2). It is often found that membranes (that is, membrane proteins) are phosphorylated on the P-side of the membrane, and glycosylated on the E-side. With respect to E-phases, P-phases are richer in protein, chemically reducing, often have a more alkaline pH and are electrically negative.

2.1.1 Membrane lipids

Most cell membranes including the double envelope of chloroplasts (but not the thylakoids) are composed mostly of phospholipids. Typical phospholipids are phosphatidyl glycerol (PG) and phosphatidyl choline (PC) (Figure 2.3). Phospholipids of this type readily form bilayer structures. The fatty acids (acyl groups) attached to the (common) glycerol group are commonly stearyl (18 carbons, no double bonds, hence denoted 18:0) and oleyl (18:1) in non-photosynthetic membranes.

Thylakoid membranes in chloroplasts contain 35% by weight of acyl lipid consisting predominantly (70%) of galactolipids such as mono-galactosyldiacylglycerol (MGDG, also known as diacylgalactosylglycerol, DGG) and digalactosyldiacylglycerol (DGDG, also known as diacyl-digalactosylglycerol, DDG). There is also a significant quantity of the sulpholipid diacylsulphoquinovosylglycerol (DSQ) and smaller proportions of PC and PG. It will be seen that all these lipids contain diacyl-glycerol (DG) (Figure 2.3). The galactolipids do not readily form bilayers, and the bilayer structure in the thylakoid membrane is presumed to depend on the protein content, because in etioplasts (Chapter 7) the protein-deficient membranes are rolled into tubes, becoming lamellar as the protein content rises.

The supposed use of specific lipids to pack proteins into the membrane, and the likely differences in the lipid compositions of the E- and P-leaflets, represent an inhomogeneity in the structure of photosynthetic membranes.

The acyl groups in thylakoids are mainly 18:1, 18:2 (olenyl), 18:3 (linolenyl), 16:0 (palmityl), trans-16:1 and 16:3. Two families of thylakoid lipids have been identified, those with 18-carbon acyl groups at position 2 of the DG moiety ('eukaryotic lipids', Roughan and Slack, 1982) and those with 16-carbon acyl groups at position 2 ('prokaryotic lipids'). PG is the only thylakoid lipid to contain mostly C-16 acyl groups at both positions 1 and 2 of its DG part. These groups have differing routes of synthesis in the envelope membranes (Chapter 7). The prokaryotic group are predominant in the thylakoids of cyanobacteria, the eukaryotic group in higher-plant chloroplast thylakoids.

Non-acyl lipids include the pigments and quinones; sterols which are important (40%) in plasma membranes are virtually absent from thylakoids (0.5%). Sterols are also absent from the cristal (inner) membrane of mitochondria, which resembles the thylakoid (see Chapter 1) and also contains a non-bilayer lipid, cardiolipin.

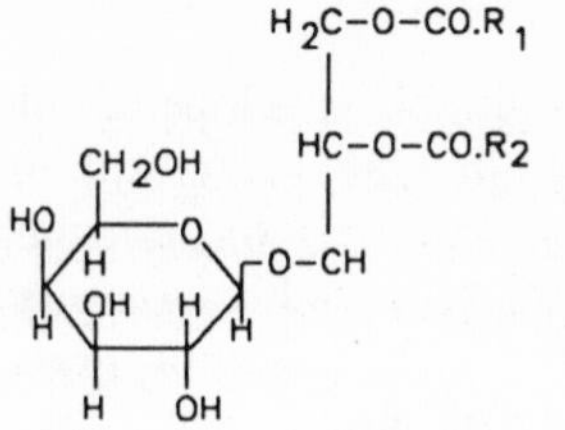

Monogalactosyldiacylglycerol

neutral

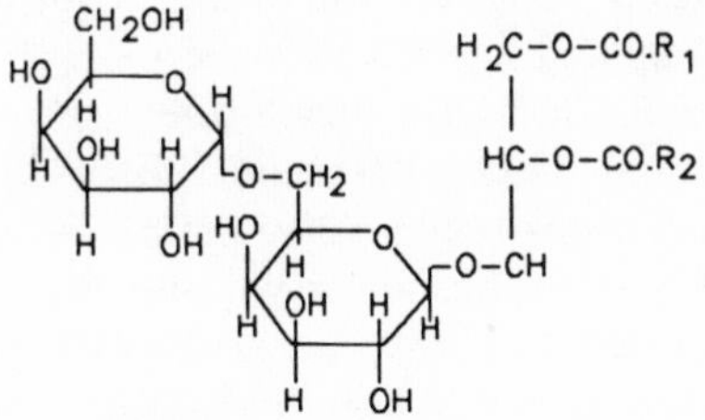

Digalactosyldiacylglycerol

neutral

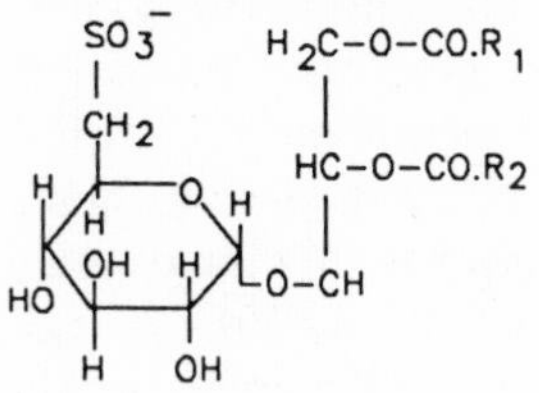

Diacylsulphoquinovosylglycerol

anionic

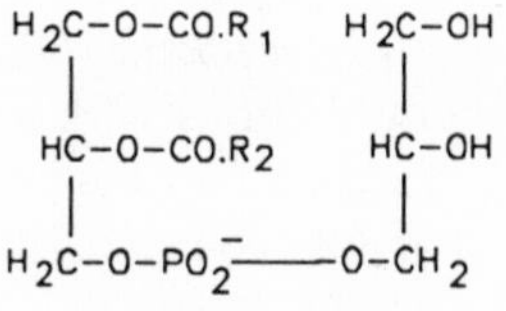

Phosphatidylglycerol

anionic

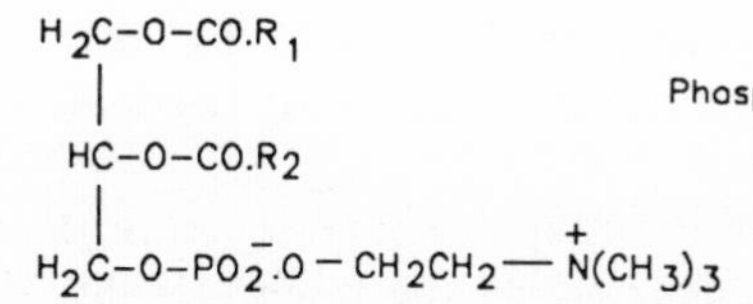

Phosphatidylcholine

dipolar ionic

Figure 2.3. Formulae of the principal thylakoid membrane lipids. The fatty acyl groups (R_1, R_2) are predominantly unsaturated members of the C16 and C18 families. Their distribution with respect to positions 1 and 2 of the glycerol are of interest (see text).

2.1.2 *Membrane proteins*

Membranes play many roles in the activity of the cell; most of these roles depend on membrane proteins and their attached groups. To summarise, we can recognise one role as *recognition and attachment*, very obvious in the attachment of cultured mammalian cells to their vessel and in cell surface antigens but also important, if less understood, in the attachment of organelles including chloroplasts to the fibrils of the cytoskeleton. The plasmalemma of higher eukaryotes is also a site where trigger substances such as hormones are recognised by and interact with receptors to cause changes in the adjacent cytoplasm. Secondly, the membranes that perform respiration and photosynthesis contain electron-transport chains, consisting of complexes of polypeptides carrying redox-active prosthetic groups. They communicate by means of small diffusible molecules: ubiquinone being lipid-soluble diffuses within the fluid structure of the lipid bilayer. Thirdly, some membrane proteins are enzymes that catalyse reactions such as the coupling of organic and inorganic reactants to the electron transport chains, or the modification of protein molecules themselves. Enzyme activities are highly specific to particular membranes. A fourth but related role is displayed by light-absorbing membrane proteins, not only those involved in photosynthesis but also rhodopsin in animal visual cells, and some mediators of phototaxes in plants. We should also note that the passage of hydrophilic solutes across a biological membrane is mediated by a set of specific carrier proteins (or complexes).

Proteins are macromolecules formed by the condensation of L-α-amino acids in unique linear sequences. The sequence of the polypeptide chain is the primary structure. The chain may contain sections with secondary structure motifs such as helix, pleated sheet or others. The chain may also fold, creating large rigid three-dimensional domains, connected by flexible hinges (tertiary structure). The side-chains of the amino acids project laterally from the polypeptide chain. Some are hydrophilic, some hydrophobic. In many cases it has been shown that water-soluble proteins arrange their hydrophilic groups on the outside. The diversity of protein in size, shape and arrangement of amino acid residues appears to be limitless.

Many membrane proteins have a quaternary structure; they are made up by the assembly of individual polypeptide molecules. Such proteins are often called particles or complexes; the individual subunit polypeptides, if they account for a characteristic property of the complex such as an enzyme activity or electron-transport intermediate, may be regarded as proteins in their own right. In this text all the subunits of complexes are referred to as polypeptides.

There are two types of multi-subunit, unpigmented protein complex that are important in photosynthetic membranes in general. The first is the quinol dehydrogenase complex which is also found in the inner mito-chondrial membrane, and in the (respiratory) membranes of bacteria. It is an essential feature of electron-transport chains, transferring electrons from quinol (ubi-, plasto- or possibly menaquinones) to cytochrome c or plastocyanin, and is described in Chapter 5. The second particle is the enzyme known as the ATP synthase (F-ATPase) that is able to couple the movement of H^+ ions, through the membrane, to the formation of ATP. The ATP synthase is even more widespread than the quinol dehydrogenase complex, since it is found in circumstances where ATP is formed from H^+ movement even where there is no electron transport (in the bacteriorhodopsin-based photosynthesis of the halobacteria). It is des-cribed in Chapter 6.

2.2   The membranes of chloroplasts and photosynthetic bacteria

Photosynthetic membranes are coloured. This is due to the presence of pigments such as chlorophyll. Although all the photosynthetic pigments are soluble in lipid solvents, they are not simply dissolved in the lipids that form the membrane (one of the old ideas). The pigment molecules are attached to proteins. In all cases analysed so far, the polypeptides with the pigments attached are complexed together, often with non-pigmented proteins, to form a set of particles (complexes) visible in the electron microscope. Such complexes commonly contain some 10% of their mass as chlorophylls and carotenoids.

Chloroplasts have a striking appearance in the electron microscope; they have a double envelope (the space between the envelope membranes being an E-space) and dense protoplasmic contents (stroma) often accounting for more than half the protein content of the cell (Figures 2.4, 2.5). The envelope membranes do not contain chlorophyll complexes: all the photochemical machinery is located in an extensive system of interconnected flattened sacs known as thylakoids ('bag-like') which appear thicker than most biological membranes, due to their high content of protein. The internal space of the thylakoid system is an E-space. The stroma contains the enzymes that account for the fixation of carbon dioxide, and the products pass out from the stroma to the cytosol via translocator proteins in the envelope. In

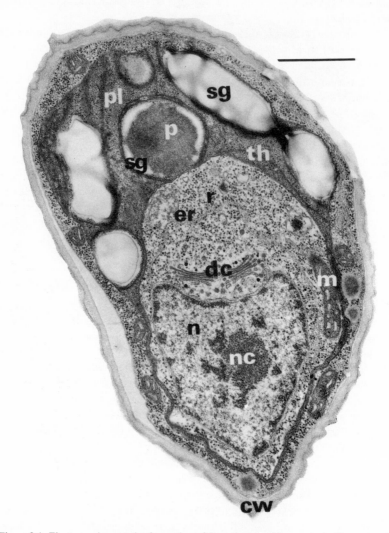

Figure 2.4. Electron micrograph of a section of *Scenedesmus obliquus*, strain D₃, mutant C-2A′. This mutant requires light for chloroplast development. The culture was grown in the dark, and this sample taken after 5 h of greening in light (white; flux rate 10 W m⁻²). The thylakoids are sparse, not fully developed and relatively unstacked. The bar represents 1 μm. Key: cw, cell wall; dc, dichyosomes (Golgi); er, endoplasmic reticulum; m, mitochondrion; n, nucleus, nc, nucleolus; p, pyrenoid; pl, plastid (developing chloroplast); r, ribosomes; sg, starch grain; th, thylakoid.

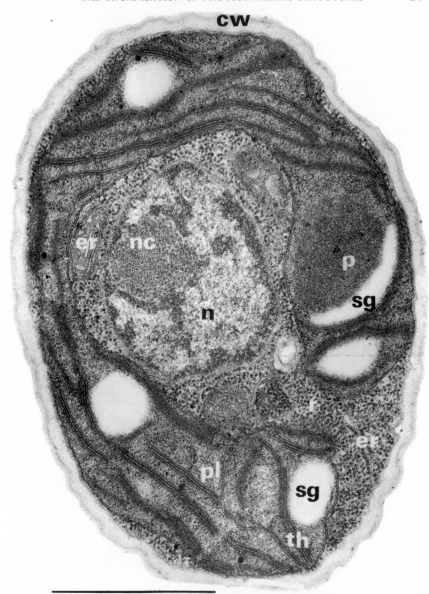

Figure 2.5. *Scenedesmus* as in Figure 2.4, later stage of greening. The thylakoids are fully mature after 7 h in light. The characteristic stacking pattern is clear. See Figure 2.4 for key. Figures 2.4 and 2.5 are unpublished micrographs by Prof. W. Wehrmeyer, Philipps-University of Marburg.

higher-plant chloroplasts the thylakoids nearly always form grana or stacks with a cylindrical appearance, connected by continuations (stroma thylakoids). Intact chloroplasts do not reveal their grana in the light microscope unless the envelope is damaged and the refractile stroma lost.

Some chloroplasts do not have grana, notably the bundle-sheath chloroplasts of some C4 plants. In algae there is more variation. The red algae show virtually no stacking (apposition); brown algae commonly show a 'banded' appearance (in cross-section) where some four thylakoids are closely appressed across more or less the whole diameter of the chloroplast.

PURPLE BACTERIA (INTRACYTOPLASMIC MEMBRANES)

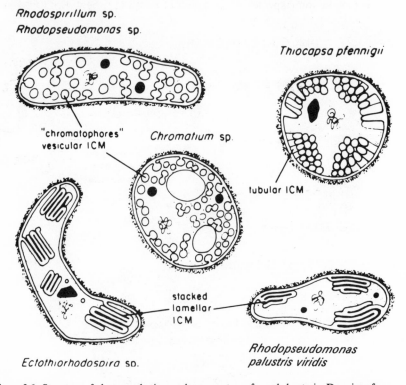

Figure 2.6. Structure of photosynthetic membrane-system of purple bacteria. Drawings from electron micrographs, showing cell wall, intracellular membrane (ICM) proliferated into chromatophores, tubular or stacked lamellar forms, DNA (represented conventionally) and polyphosphate deposits (dark lumps). Note that in some species the form of the ICM depends on the light-level of the culture. From Sprague and Varga (1986).

The green algae and the euglenoids are similar to higher plants in that they possess the granal structure.

Thylakoids occur throughout the cells of *Prochloron* (the best-described member of the prochlorophyta), and of the cyanobacteria; there is no chloroplast envelope. The thylakoids of cyanobacteria (like those of red algal chloroplasts) are not usually appressed and in some cases they are branched. *Prochloron* on the other hand shows thylakoid stacking: stacking is correlated with possession of chlorophyll b.

In purple and green photosynthetic bacteria there is only one membrane, the cell membrane (CM) but in most species of purple bacteria it becomes highly invaginated (intracytoplasmic membrane, ICM) (Figure 2.6). The extended membrane surface forms a system of lamellae resembling thylakoids in some species, or a system of tubes in others. Both CM and

GREEN BACTERIA (CHLOROSOMES)

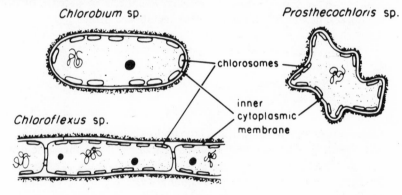

Chlorobium sp.   *Prosthecochloris* sp.

chlorosomes

inner
cytoplasmic
membrane

Chloroflexus sp.

SIMPLE PHOTOSYNTHETIC BACTERIA
(NO CHLOROSOMES OR INTERNAL MEMBRANES)

Rhodospirillum tenue
Heliobacterium chlorum

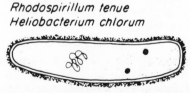

Figure 2.7. Structure of photosynthetic membrane system of green bacteria. Drawings from electron micrographs, showing cell wall, cell membrane, chlorosomes (and their absence from 'simple' cells), DNA (represented conventionally) and polyphosphate deposits (dark lumps). From Sprague and Varga (1986).

ICM have been shown to carry pigmented protein complexes and to be active in photosynthesis, and it may be assumed that the formation of ICM is an adaptation to increase the light-absorption of the cell. In the green bacteria there is no ICM (Figure 2.7); the cell membrane carries on its inner surface characteristic assemblies of pigmented protein known as chlorosomes. Chlorosomes are often lost in the course of preparation of sections for electron microscopy, giving rise to the holes known as 'Chlorobium vesicles'.

Heliobacterium chlorum is a recently discovered photosynthetic bacterium (Gest and Favinger, 1983) that is green in colour, but does not have chlorosomes or ICM. The membrane is unusually rich in protein, and the principal pigment is bacteriochlorophyll g.

Halobacterium halobium as a member of the archaebacteria has distinct types of lipid in its membrane. The membrane normally contains the electron-transport chain and F-ATPase proteins associated with respiration; when bacteriorhodopsin develops it collects in patches of purple membrane in which there is no other membrane protein.

2.3 Pigments

2.3.1 *Spectroscopy*

In the next section we shall discuss the photosynthetic pigments. A pigment is a chemical substance that absorbs visible light and therefore appears coloured, accounting for the natural or artificial colours of objects. Often the light-absorbing property of a pigment molecule can be identified with one part of its structure, and that part is then known as the chromophore. Chromophores commonly have either a conjugated structure of alternating double and single bonds, or unpaired electrons.

The absorption of light is technically very simple to measure precisely, and usually does not damage the sample. It is usual to deal with absorption, rather than transmission, because of a useful relationship known as the Beer–Lambert law. If a monochromatic beam of light (parallel or collimated) is passed through a solution of an absorbing material, the emerging beam will have an intensity I that is less than the original intensity I_o. The quantity known as the absorbance of the solution, A, is given by:

$$A = \log_{10}\frac{I_o}{I} = A_M.c.l$$

where l is the path length (in cm), c is the concentration in molar units $(mol\,dm^{-3})$ and A_M is a constant for the particular substance and wavelength, known as the molar absorption coefficient. An alternative formulation uses the term extinction (E). The measurement of absorption (the logarithmic calculation is performed by the instrument, the spectro-photometer) provides a non-destructive means of determining concentration. The Beer–Lambert law applies to most solutions, provided allowance is made for chemical or photochemical reactions that may be taking place, but fails with suspensions when self-shading or light-scattering occurs. Chloroplasts and their grana deviate from the law, but extracted pigments and pigment–protein complexes, and even fragments of membrane containing the complexes, obey it satisfactorily. Deviation from the Beer–Lambert law is also observed when the measuring beam is insufficiently monochromatic. A useful rule is that the spectral width of the beam should be one-fifth or less of the half-amplitude width of the spectral absorbance peak being measured. The sharpness of the peaks of chlorophylls necessitates a spectral bandwidth of 2 nm or less, readily achievable by means of prism or grating monochromators, but not with colour filters.

When absorption is measured and plotted as a function of wavelength, the resulting curve (the absorption spectrum) is characteristic of the material and may be used for identification. When absorption is measured as a function of time, it provides the means of detecting and measuring the speed of chemical reactions and electronic changes that take place in times down to one picosecond (1 ps $= 10^{-12}$ s).

Many other materials in the cell are intrinsically coloured without occurring in sufficient concentration to be recognised as pigments. Still more can be observed spectroscopically by extending the range of wavelengths into the shorter ultraviolet (down to 180 nm, or 160 nm if oxygen is excluded) or into the infrared.

2.3.2 *Chlorophylls, phycobilins and carotenoids*

The photosynthetic pigments are of three kinds, chlorophylls, carotenoids and bilins. Chlorophylls and bilins are both tetrapyrroles, the former cyclic, the latter linear. Chlorophylls (Figure 2.8a) contain magnesium coordinated inside the chlorin ring system. Chlorophyll differs from haem in the side-chains (the acid groups have been removed or esterified, rendering the molecule very hydrophobic, and there is an isocyclic ring) and in the saturation of ring IV in chlorophyll (Chl), rings II and IV in bacteriochloro-phyll (Bchl) (Figure 2.8b).

Chlorophylls

(a)

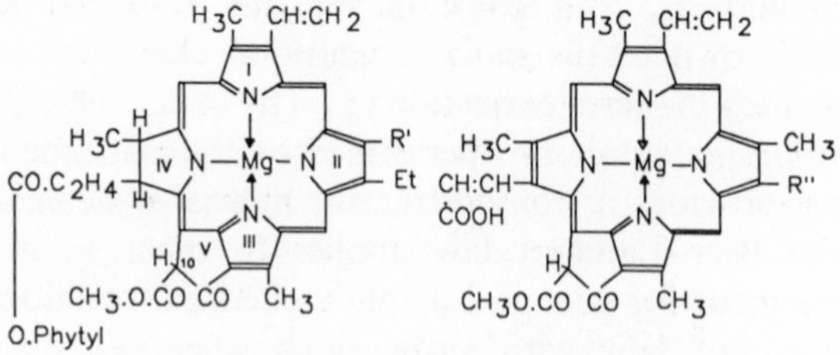

R' = CH₃ in Chl a
   = CHO in Chl b
R'' = Et in Chl c₁
   = CH:CH₂ in Chl c₂

Phytol:

Bacteriochlorophylls

(b)

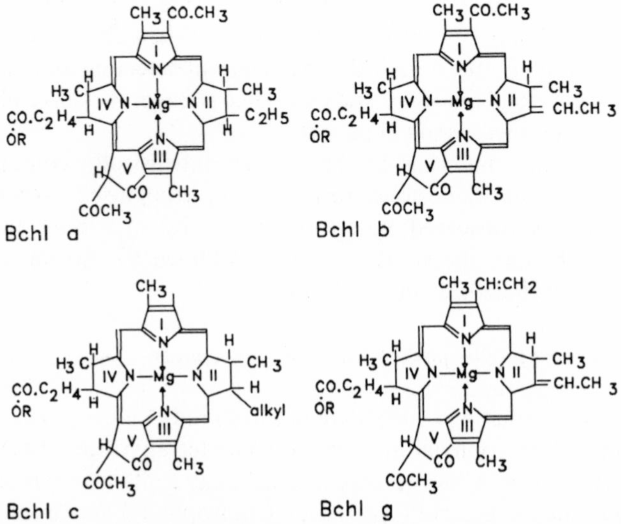

Bchl a   Bchl b

Bchl c   Bchl g

R = phytyl or

farnesyl:

So far as is known, pigments are always attached to protein in photosynthetic structures. Chlorophylls are coordinated via the magnesium atom to histidine residues (or other amino acids) in the protein, and are easily detached by solvent extraction.

The phycobilins or linear tetrapyrroles (Figure 2.9) are derived from cyclic tetrapyrroles by oxidation and removal of a methene bridge. Bilins are attached to protein by covalent bonds to the sulphur of cysteine residues, forming the phycobiliproteins. These are water-soluble, and readily detached from the thylakoid membranes, for example by freezing and thawing. Four classes have been recognised (see Zuber *et al.*, 1987, for a review): allophycocyanin (APC, absorption maximum at 650–680 nm), phycocyanin (PC, 620–635 nm), phycoerythrocyanin (PEC, 575 nm) and

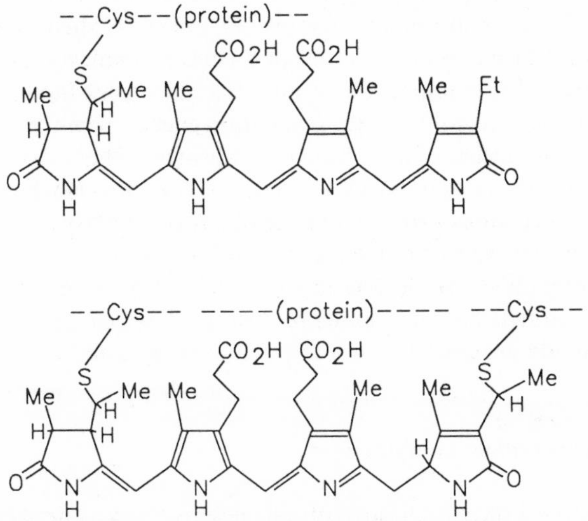

Figure 2.9. Linear tetrapyrroles: formulae of phycobilins. PC, phycocyanobilin (upper) and PE, phycoerythrobilin (lower), linked to protein by one and two thioether bonds to cysteine, respectively. After Glazer (1984).

Figure 2.8. Cyclic tetrapyrroles: formulae of (a) chlorophylls a, b and c, and (b) bacteriochlorophylls a, b, c and g. R is phytyl in Bchl a and b, and farnesyl in Bchl c and g. All the formulae are oriented with the *y*-axis vertical and the *x*-axis horizontal. Note that all the pigments shown have the isocyclic ring V, but Chl c does not have the saturation in ring IV that defines true chlorophyll. The hydrogen at C-10 (in ring V) is directed away from reader, except in the controversial epimer of Chl a, Chl a′.

phycoerythrin (PE, 545–565 nm). Phycobiliproteins obtained from red algae are prefixed R- and those from cyanobacteria C-, thus R-PE, C-PC, etc. They are the principal determinant of the colour of these cells, in which they may account for some 50% of the protein. When detached from the membranes, they show an intense fluorescence.

The protein subunits to which the phycobilins are attached have masses of 18–22 kDa (α, β in cyanobacteria, and in addition one of about 30 kDa, γ, in red algae). Each polypeptide carries usually one but up to three phycobilin units. They join in quaternary structure, usually the hetero-hexamer $(\alpha\beta)_6$, making discs, which are connected together in the phyco-bilisome (see below).

The carotenes are hydrocarbons, usually C_{40}, and the xanthophylls (Figure 2.10) are derived from the carotenes by addition of oxygen as hydroxyl, oxo or epoxy groups. The term 'carotenoid' covers both. Xanthophylls occur in antenna complexes, and pass their energy to the chlorophylls. Carotenes occur in reaction-centre complexes, and probably do the same, but it is important to note that carotenes have a protective role. They are able to collect excess excitation energy from the chlorophylls after very bright flashes of light have been absorbed (the 'valve-reaction' of H.T. Witt, probably not important in nature). They are also able to provide a sink for triplet states produced in the chlorophyll mass (see Chapter 3). Triplet states can react with oxygen, and damage to the system would ensue. Carotenoids are effective in sacrificially preventing such damage, and it is observed that they undergo a cycle in which, in the light, they acquire epoxide groups, subsequently removed by a reductive enzyme.

2.4 Pigment–protein complexes

From the earliest days of photosynthesis research, it was debated whether chlorophyll was dissolved in lipid or attached to protein. The first indication of the presence of protein complexes came from the work of Smith (1938) who showed that an extract of leaves made by means of the surface-active agent digitonin contained chlorophylls attached to a protein fraction. In the fifty years that have followed, the variety of chlorophyll-proteins has become more apparent, and the use of more gentle procedures has diminished the quantity of unbound chlorophyll to the point where it is clear that all chlorophyll is attached to proteins, the proteins being 'dissolved' in the lipid bilayer.

Functionally, there are two kinds of pigmented protein: antenna (found

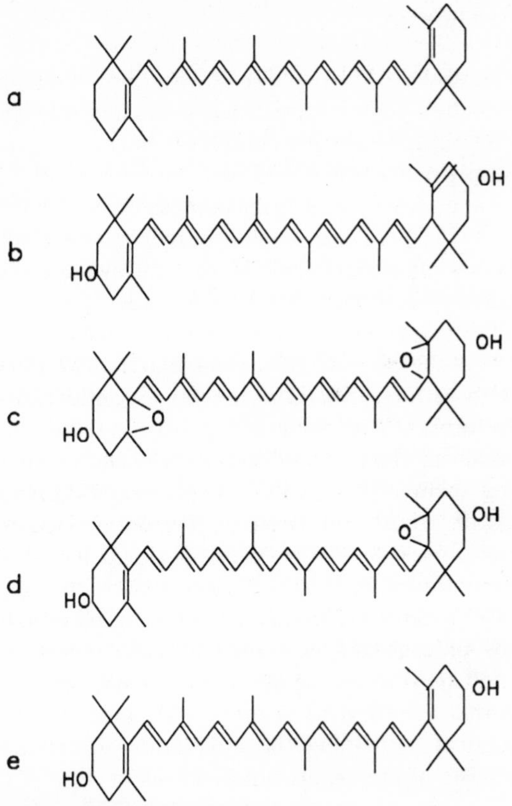

Figure 2.10. Formulae of some carotenoids. Note the isoprenoid structure (repetition of a branched C-5 group) and the conjugated double bonds that generate the visible colour. β carotene (*a*) is the principle carotene of green plants and is found in reaction centre complexes; the principal xanthophylls lutein (*b* the principal form in higher plants), violaxanthin (*c*), antheraxanthin (*d*) and zeaxanthin (*e*) are found in antenna complexes. The last three undergo interconversion in the protective xanthophyll cycle.

in light-harvesting complexes) and reaction-centre complexes. Reaction centres are often associated with 'core antennas' making a 'core complex' (see Figure 2.14). There is much variety in the nature of antenna complexes, but there appears to be much more uniformity in reaction-centre complexes; perhaps there are as few as three or even two types, as judged by the apparent homologies in the amino-acid sequences of the essential proteins.

2.5 Antenna complexes

Antenna complexes have the function of absorbing light, and passing the energy to the reaction-centre complexes (the mechanism is described in the next chapter). The reaction-centre complexes, of course, contain their own complement of pigments, and could perform photosynthesis without the aid of the antennae. However, antenna complexes are needed because even in full sunlight, a single molecule of chlorophyll or bacteriochlorophyll can only absorb a few photons in one second on average. Reaction-centre complexes are able to carry out chemical reactions at a much greater rate, of the order of hundreds of molecules per reaction centre per second (limited by the diffusion of the reactants and products). Furthermore, the average intensity of illumination throughout the day is considerably less than full sunlight, and photosynthetic bacteria live on or in pond mud at still lower levels of illumination. The turnover rate of the reaction-centres is increased by surrounding them with antenna complexes, containing in total, per reaction centre, fifty to several thousand pigment molecules. The antenna pigments not only provide a bigger cross-sectional area for the capture of light, but also capture a bigger proportion of the visible range of wavelengths. This is achieved because antenna complexes contain a greater range of types of each pigment such as carotenoids, phycobilins or chlorophylls, and also modify the physical environment for these pigments in such a way that there are significant variations in the positions of the absorption maxima, spreading the absorption spectrum of the whole complex over more of the visible range. This widening of the absorption spectra is most noticeable in the antenna complexes of purple bacteria, affecting the pigment bacteriochlorophyll, and in those of (for example) the Chromophycean algae, where carotenoids such as fucoxanthin are chiefly affected. Red-shifts of up to 100 nm have been observed, with respect to the same pigments when extracted.

One may suspect that in a given species the proportion of pigment contained in the antenna complexes is adapted to the average light intensity and probably spectral composition as well. Moreover, in several cyano-bacteria the C-PCs are variable in response to changes in the light regime, a phenomenon known as chromatic adaptation (Bogorad, 1975).

In two cases antenna complexes are not in the membrane, but are attached to it on the cytoplasmic (or stroma) side. These are the chlorosomes of green bacteria, and the phycobilisomes of cyanobacteria and red algae.

2.5.1   Chlorosomes

Chlorosomes (Figure 2.11) are large arrays of pigment on the inside surface of the cell membrane, thus they project into the cytoplasm. Because in early studies the contents were dissolved away, chlorosomes used to be called chlorobium vesicles. They are found in both groups of green bacteria (sulphur and non-sulphur), and are the main common feature of these two groups. The chlorosome is based on a protein molecule which may be regarded as a cylinder on which are placed seven Bchl c (or d or e) molecules absorbing at 740 nm, and the cylinders are grouped in aggregates containing twelve protein subunits and hence 84 Bchl c molecules. These aggregates are aggregated further so as to form the chlorosome containing some 10 000 Bchl c molecules, and there may be 100 chlorosomes in a cell. The groups of cylindrical subunits are connected to a baseplate protein, which also carries Bchl c, and at the point of attachment to the membrane there is a linker protein. In the case of *P. aestuarii* this linker protein, carrying seven Bchl c molecules, has been crystallised and its tertiary structure determined. This demonstrated that the pigment molecules are rigidly and precisely fixed to the protein chain, with nearest-neighbour spacing of 1.2 nm within one subunit and 2.4 nm between pigments in adjacent subunits. The nearest-neighbour orientations are of the order of 40°. The details are important for understanding the high efficiency of the energy transfer to the reaction-centre complex, which is in the membrane. It is expected that when similar studies are completed for other antenna proteins the same spacing and orientation rules will be found to apply.

2.5.2   Phycobilisomes

Phycobilisomes (Figure 2.12) are attached to thylakoids, in the cells of cyanobacteria and the chloroplasts of the algal groups Rhodophyta and Chromophyta. This common structural feature is one reason for suspecting an evolutionary link between cyanobacteria and chloroplasts. Phyco-bilisomes project into the stroma, and are similar to chlorosomes insofar as there are pigmented protein subunits that are arranged in a hierarchy of aggregate structures, and are connected to the membrane by linker proteins. In this case, however, there is a more obvious gradation of pigment types. Only phycobiliprotein pigments are present, as hexameric $(\alpha\beta)_6$ double-discs described above. The discs are assembled on a central unpigmented protein rod, in such a way that a thick stack is formed. The

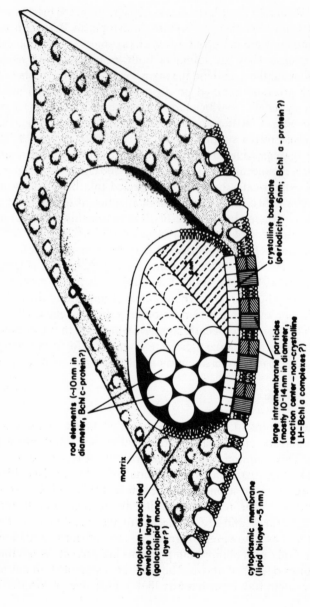

Figure 2.11. Diagrammatic structure of the chlorosome of a green sulphur bacterium *Chlorobium limicola*, based on the freeze fracture micrographs of Staehelin *et al.* (1980).

rod elements (~10nm in diameter, Bchl c-protein?)

matrix

cytoplasm-associated envelope layer (galactolipid mono-layer?)

cytoplasmic membrane (lipid bilayer ~5 nm)

large intramembrane particles (mostly 10-14 nm in diameter, reaction center-non-crystalline LH-Bchl a complexes?)

crystalline baseplate (periodicity ~ 6nm; Bchl a - protein?)

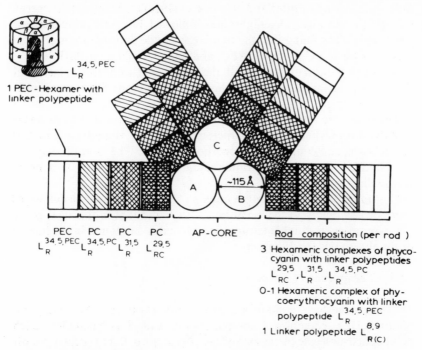

Figure 2.12. Diagrammatic structure of a phycobilisome from a cyanobacterium, *Mastigocladus laminosus*. Three rods form a core, parallel to the membrane; they contain allophycocyanin (AP) trimers assembled round a linker polypeptide. Radiating from the core are rods made up of discs; the discs contain phycocyanin (PC) hexamers around linker polypeptides (L_R) except for the outer ends which are made of phycoerythrocyanin (PEC). The PC discs are non-identical in absorption. The shading indicates a progression towards longer wavelengths. There is no phycoerythrin in this phycobilisome. From Zuber *et al.* (1987).

phycobilins at the distal end of the stack are either PE or PEC (appearing red in colour, absorbing blue light, that is photons of the higher energy range). The bulk of the stack is PC. The linker proteins affect the pigment binding in such a way that the wavelengths of maximum absorption are adjusted or tuned and form a series progressing towards the red (lower energy of excitation) at the proximal end: PEC (575 nm) > PC (620–635 nm) > APC (650–655 nm) > Chl a (670–680 nm). The structure of the phycobilisome of *Mastigocladus laminosus* has been worked out in detail (Figure 2.12, see Zuber *et al.*, 1987) and the energy flow has been mapped for all pairs of chromophores. The technique of time-resolved fluorescence, in which the changes of fluorescence emission are followed in the 10 ps to 1 ns time scale, allowed Porter and co-workers (1978) to establish that energy transfer (in red algal phycobilisomes) followed Förster's principles (see Chapter 3). They found that the time to reach the fluorescence maximum was 12 ps for R-PC, 24 ps for APC, and 50 ps for Chl a (in PSII).

Although they are unpigmented, the amino-acid sequences suggest that the linker proteins are related to the phycobiliproteins. Elsewhere, linker proteins are pigmented.

The stacks are attached at right angles to a core formed of three stacks of APC discs (cylinders). The structure of the core is analogous to that of the rods but contains different polypeptides. The overall shape of the phycobilisome may be hemidiscoidal (the example above) or hemi-ellipsoidal.

Contrary to the principle that antenna structures should be connected more or less rigidly to the membrane, it is found that in the algal group of Cryptomonads, soluble tetrameric phycobilins are found in the lumina of the thylakoids (an E-space).

2.5.3   The antenna of the purple bacteria

The purple bacteria contain antenna complexes (light-harvesting complexes, LHC) assembled from pairs of α and β polypeptides. Each polypeptide molecule carries one or two molecules of Bchl a (Bchl a in most species, Bchl b in some such as *Rps. viridis*) (Figure 2.13). The various antenna complexes of the purple bacteria can be systematised into three groups as shown in Table 2.1 It may be noted that all species have one antenna of a long-wavelength form, while some others possess a mid-range second antenna, and some of those a third, shorter-wavelength form. There are some 40 Bchl in antenna complexes for every reaction centre in the chromatophore membranes of (for example) *Rb. spheroides*. 25–30 of these are in B875 (see Table 2.1) in a fixed proportion; a variable amount of B800-B850 accounts for the remainder. It is likely that the antenna structure may be rearrangable in the membrane so as to adjust for varying intensity of illumination.

The green non-sulphur bacterium *Chloroflexus aurantiacus* possesses an antenna corresponding to the B800-B850 type. It is connected to the chlorosome by means of a pigmented linker protein containing bacterio-chlorophyll a (790 nm, Betti *et al.*, 1982). The principle of progression of increasing wavelengths of absorption maxima from outer antenna to reaction centre (865 nm) is clearly illustrated.

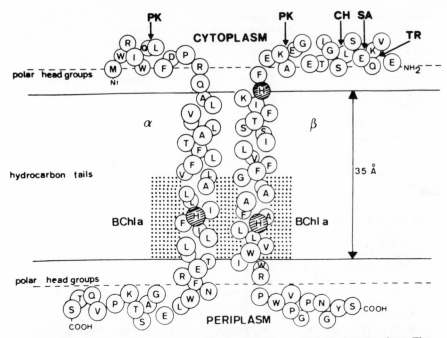

Figure 2.13. One unit of the antenna (B880) protein from a purple bacterium, *R. rubrum*. The two polypeptide chains (α, β) span the membrane, and carry two bacteriochlorophyll molecules in the shaded region attached to the histidine residues (H). A third Bchl may attach to the histidine residue shown at the cytoplasmic face of the β-subunit. Amino acids are identified by the standard one-letter code. The symbols PK, CH, SA and TR indicate the sites of attack by the proteinase K, chymotrypsin, *S. aureus* proteinase and trypsin, respectively, which identified the N-terminal ends as being on the cytoplasmic face of the membrane. From Zuber *et al.* (1987).

2.5.4   Complexes in the green plant chloroplast

There are at least twelve chlorophyll-carrying polypeptides in chloroplasts. They are organised into three particles that can be seen in freeze-fracture electron micrographs, and isolated by chromatography or electrophoresis following extraction by means of detergents. They are known as PSI and PSII (which retain photochemical activity and therefore contain reaction centres) and LHCII, which is only an antenna, principally connected to PSII. Terms may become confusing when the particles are further subdivided. Both PSI and PSII contain many (234 and 34) chlorophyll

Table 2.1 Antenna complexes of the purple bacteria.

Antenna type	B890	B875	B800-850 Type I	B800-850 Type II	B800-820
			Bacteriochlorophyll spectral ranges		
Bchl a (mol)	2	2	2	3	3
carotenoid (mol)	1	2	1	1	1
polypeptides	2	2	2–3	2	2
amino acids		52–58,	54–60,	Approx.	Approx.
residues in pp.	52, 54	47–48	52	50–65	50–65
Examples:					
R. rubrum	+				
Rps. viridis*	+				
Rps. palustris		+	+		
Rb. capsulatus		+	+		
Rb. sphaeroides		+	+		
Rc. gelatinosus		+	+		
C. vinosum	+			+	+
Rps. acidophila					
7050, high light		+		+	+
7050, low light		+			+

Note: The quantitative data are based on analyses of complexes isolated from only a few of the species given as examples, and represent 'minimal compositional units'. Data from Thornber et al. (1983).

*Rps. viridis, possessing Bchl b, has an antenna absorbing maximally at 1015 nm; the two polypeptide subunits show strong homology to those of R. rubrum.

molecules, attached to several polypeptides. PSI has been resolved by more severe detergent treatment into an active particle, RCI (PSI core), and two antenna polypeptides LHCIα and LHCIβ (Haworth et al., 1983; Lam et al., 1984). LHCI (α and β) do not have the independent, detachable, existence that LHCII has. RCI has some 90 chlorophylls, and since only some three are needed to explain the photochemistry, the rest are described as antennas (peripheral, internal and core, describing the progressive removal of chlorophyll from RCI by increasingly severe detergent treatment). One must therefore distinguish antenna chlorophylls in RC particles from antenna particles themselves. It is interesting however that all the LHC polypeptides (except probably LHCIIa) contain chlorophyll b (in higher plant chloroplasts), as well as the xanthophylls. The core particles PSII and RCI contain only chlorophyll a and carotenes.

In PSII on the other hand most of the chlorophyll can be removed attached to two polypeptides leaving a reaction centre RCII containing only (probably) six pigment molecules. There is no convenient name for the PSII core-antenna peptides, apart from the trivial CP (chlorophyll-protein) nomenclature based on the apparent molecular masses observed when the

Table 2.2 Chlorophyll-protein complexes in higher plants.

Complex	Symbols	Polypeptides Mol mass		Gene		Pigments			
		PAGE	Abs.	Sym.	c/n	Chla	Chlb	Car	Xan
		(kda)							
Photosystem I									
Core	CCI, CPI, IA	60–70	83.2	psaA	c	90	0	2	0
	IB	60–70	82.5	psaB	c	+P700			
Antenna	LHCIa	22, 23*	24	Cab	n	a/b = 3.5–3.7		0	+
	LHCPIb	20, 21*	24	Cab	n				
Total antenna						114	30		
Total PSI						204	30		
Photosystem II									
Linker	LHCIIa, CP29	29	N?	—	n?	+	0	0	+
Antenna	LHCP, CPII, LHCIIb	25	25.2	Cab	n	7	6	0	3
	LHCIIc, CP27	27		Cab	n	+	+	0	+
	LHCIId, CP24	24				a/b = 1.25		0	+
Total antenna						129	79		
Core									
Int. Ant.	CPa-1, CP47	47	56.2	PsbB	c	30	0		
	CPa-2, CP43	43	51.8	PsbC	c				
R.C.	D_1	32	37.2	PsbA	c	P680			
	D_2	32	39.4	PsbD	c	4–5		1	
Total core						34	0		
Total PSII + LHCII						163	79		

Sources: Thornber (1986, 1988), Peter and Thornber (1988), Green (1988) and Dyer (1988).
*At least one polypeptide in each group carries chlorophyll.
For each pigmented component of each complex, synonymous symbols are listed, followed by values of the molecular mass as estimated by SDS-polyacrylamide gel electrophoresis (PAGE) and by summing the amino acid composition determined from the gene sequence (Abs.). Gene symbols correspond with Figure 7.11 for chloroplast (c) location; (n) denotes nuclear coding.

green polypeptides are run in polyacrylamide-gel electrophoresis in the presence of sodium dodecyl sulphate (SDS) thus CP43 for an apparent 43 kDa. It will be noted that many polypeptides are allocated symbols based on SDS-electrophoretic mobility; most biophysical techniques for accurate determination of relative molecular mass require too much purified material, and are invalidated by the detergents usually used to extract the material. However, in many cases the genes that code for the polypeptides have been identified and their sequences determined, which provides an accurate molecular mass assuming that the pigment complement, and the extent of post-transcriptional modification of the gene product, are known. These relationships are set out in Table 2.2.

LHCII contains nearly half of the total chlorophyll in the thylakoid. It can be resolved into three polypeptides, LHCIIB, c and d. There is a fourth polypeptide, LHCIIa, which at present seems to be a linker protein, probably belonging to, but easily detached from, PSII. All these polypeptides are coded by genes in the nucleus of higher plants, and it appears that there are several non-identical versions of each gene present (a problem reviewed by Chitnis and Thornber, 1988). This complicated variability makes the biochemical study of LHCII difficult.

LHCII (in total) contains most of the chlorophyll b of the chloroplast, a more or less equal quantity of chlorophyll a and xanthophylls, chiefly lutein, violaxanthin and neoxanthin. It is associated with PSII (the majority form) and confined to the grana (the appressed thylakoid membranes). It may undergo phosphorylation and move out of the grana in a regulatory fashion (see Chapter 7).

LHC complexes have been isolated from many different algal groups. The Chlorophyta generally contain normal LHCI and LHCII, apart from the substitution of siphonoxanthin for lutein in the Siphonales. The Chromophyta on the other hand possess complexes which are smaller (13–17 kDa) than LHCI or II, and show no homologies. Chlorophyll c (rather than b, Figure 2.8a) accompanies chlorophyll a, and the xantho-phyll may be fucoxanthin (in complexes usually coupled to PSII) or violaxanthin (coupled to PSI, in brown algae). Fluorescence excitation spectra show that the xanthophyll is efficiently connected to chlorophyll a. The Cryptophyta also contain chlorophyll c_2, and water-soluble complexes containing the xanthophyll peridinin have been obtained from dinoflagel-lates. (See Anderson and Barrett, 1986 for a review.)

The possession of LHCII complexes, of whatever sort, is correlated with stacking, or the tendency of thylakoids to appress themselves into groups. Chloroplasts with chlorophylls a and b tend to form stacks, up to six thylakoids deep in the Chlorophyta (variable) leading to the very uniform grana found in higher plants. Chlorophyll c-types tend to form groups of three appressed thylakoids.

Prochloron is one of two known species of Chlorophyta, which like the Cyanobacteria, and unlike all other photosynthetic prokaryotes, contain chlorophylls a and b, and evolve oxygen. *Prochloron didemnii* was discovered as a symbiont in tropical didemnid ascidians (sea squirts). A second free living member has recently been found. *Prochloron* contains LHCI and LHCII complexes (Table 2.3). It shows some degree of appression of its thylakoids, although most are loose. The LHCII contains less chlorophyll b in *Prochloron* than in the Chlorophyta, the protein is larger (34 as opposed to 25–29 kDa) and there is no immunochemical cross-reactivity; cross-reactivity is found between PSI and PSII poly-peptides of all groups.

Table 2.3 Pigment composition of green plant antennas.

Group	PS II antenna		PS I antenna	
	Chl	Principal xanthophyll	Chl	Principal xanthophyll
Cyanobacteria: Free living	Phycobilisome	No intrinsic antenna known	None known	
Symbiotic (*Prochloron*)	a/b: 2.3	Zeaxanthin	(a + b)	N.I.
Green algae and	LHCII:		LHCI:	
higher plants	a/b: 1.0–1.4	Lutein	a/b: 3.5–3.7	N.I.
(N.B. Siphonales)	a/b: 0.66	Siphonoxanthin	a/b: 1.8	Siphonoxanthin
Phaeophyta	a/c_2:2	Fucoxanthin	$a/(c_1 + c_2)$: 3	Violaxanthin

N.I.: not identified. Data from compilation by Anderson, J.M. and Barrett, J. (1986) in Staehel and Arntzen, eds, *Photosynthesis III, Encyclopaedia of Plant Physiology*, new series, vol. 19, Springer-Verlag, Berlin, pp. 269–285.

2.6 Reaction-centre complexes

2.6.1 *Green plants possess two types of photosystem*

In the early 1960s it became clear that green-plant photosynthesis could not be explained on the basis of a single type of photosystem. Owing to the two photosystems having different absorption spectra, the effect of light of say 700 nm wavelength was discovered to be different from that of light of 680 nm. For example, the chloroplast cytochrome f was reduced (shown by the increase in its α-band at 554.5 nm) by illuminating a leaf with 680 nm light, and oxidised by 700 nm light. Also, the intensity of fluorescence was decreased by prior illumination with 700 nm light and increased by 680 nm light. The two light effects were numbered I and II by L.N.M. Duysens, and these labels have persisted; photosystem I preferentially absorbs 700 nm light.

Both wavelengths were required simultaneously for the most efficient utilisation of light energy. This was shown by the phenomenon of *enhancement*: two light sources are arranged, one at 680 nm and the other at 700 nm. The rate of photosynthesis with both wavelengths together is greater than the sum of the rates with each wavelength alone (R. Emerson). In an alternative demonstration, it was shown that the rate of photo-synthesis greatly increased, temporarily, when illumination at 700 nm was changed to 680 nm (the *chromatic transient*). The explanation for these effects was that normal green-plant photosynthesis depended on two light reactions in sequence; cytochrome *f*, and the determinant of fluorescence, are located between the photosystems and hence are reduced by II and oxidised by I (see Chapter 4).

The historical details, shown in the time chart (Table 2.4), are perhaps academic; the isolation of two different photosystem particles from chloroplasts provides the modern basis for the theory of photosynthesis. Nevertheless, it should be borne in mind that of the two photosystems, PSI and PSII, PSII is unable to make efficient use of light of longer wavelength than 680 nm.

2.6.2   *Purple bacteria possess the simplest reaction centre*

Reaction-centre complexes are of at least two types. One type, found in the purple bacteria (PBRC), was isolated by Reed and Clayton (1968) and studied intensively. PBRC preparations have been obtained from many

Table 2.4 Time Chart: evidence for two light reactions in green plants.

Date	Workers	Observation	Inference
1932	R. Emerson W. Arnold	Photosynthetic unit = 2400 Chl/O_2	Either most Chl is inactive or Chl cooperates
1941	Emerson and C.M. Lewis	Quantum requirement for O_2 production = 8 to 10	Most Chl not inactive
1943	Emerson and Lewis	'Red Drop': quantum requirement rises steeply above 680 nm	Far-red absorbing Chl separate function from 680-form
1957	Emerson *et al.*	'Enhancement' (see text)	Both functions required for O_2 production
1957	L.R. Blinks	'Chromatic Transients'	Product of far-red is substrate for red absorbing Chl
1957	B. Kok	Photobleachable pigment absorbing at 700 nm	Identify active Chl of far-red Chl form
1953–6	L.N.M. Duysens	Different effects of light-I and light-II on initial fluorescence level	Define photosystems I and II
1960	R. Hill and F. Bendall	Redox potentials of cyts *f* and b_6 are intermediate	*if* cyt. *f* is involved then two light reactions must exist between O_2/H_2O and $NADP^+/NADPH$
1963	Duysens	Lights I and II affect cyt. *f* redox state differently	Cyt. *f* is involved
1964	N.K. Boardman and J.M. Anderson	PSI and PSII fractionated by means of digitonin	
1969	G. Döring *et al.* (H.T. Witt's lab)	Observation of Chl a_{II}	Active Chl of PSII (= P680)

species, representing both the Rhodospirillaceae (formerly Athiorho-daceae) and Chromatiaceae (formerly Thiorhodaceae). In addition, the reaction centre of the species *Chloroflexus aurantiacus* (green non-sulphur bacteria) proved to belong to the group. The details are summarised in Table 2.5. Typically, a PBRC contains four Bchl a molecules and two bacteriophaeophytin (Bph) molecules (3 Bchl and 3 Bph in *Chloroflexus*) in addition to one carotenoid molecule. In most cases a quinone molecule (ubiquinone in *Rb. sphaeroides*, menaquinone in *Rps. viridis*, see Figure 4.3) is present, together with an iron ion.

The reaction centres from the latter two species have been crystallised and their three-dimensional structure analysed by X-ray crystallography (see Deisenhofer *et al.*, 1984, for *Rps. viridis* and Allen *et al.*, 1987 for *Rb. spheroides*). The student is strongly urged to examine their coloured stereograms: the pigments are carried by two polypeptides, which they connect, known as L (light) and M (medium); in the purple bacteria there is an unpigmented H (heavy) polypeptide, not found in the green non-sulphur bacterium *Chloroflexus*. Some purple bacteria also have a c-type cyto-chrome with four haem groups (no pigment). The operation of reaction centres is described in the next chapter.

Table 2.5 Comparative composition of purple bacterial reaction centres.

Species	Rb. sphaeroides	Rps. viridis	Chloroflexus aurantiacus
Pigments (mol):			
Photoactive centre	P870	P960	P865
Bacteriochlorophyll	4 Bchl a	4 Bchl b	3 Bchl a
Bacteriophaeophytin	2 Bph a	2 Bph b	3 Bph a
Carotenoid	1	1	0
Quinone*:	$2\ UQ_{10}$	$1\ MQ_7, 1\ UQ_{10}$	2 MQ
Polypeptides (PAGE-mass):			
H subunit	28	35	—
M subunit	24	28	30
L subunit	21	24	28
C-cytochrome	—	38 (4 haems)	—

*Not all the quinone complement quoted above is found in the crystal structure; presumably it is lost during manipulation. The redox intermediates P870 (etc), quinones and cytochrome are treated in Chapter 3.
Source: Thornber *et al.* (1983) (adapted). *C. vinosum* (Figure 3.4) appears to be similar to *Rps. viridis* in overall composition, except that it has Bchl a instead of Bchl b, and values are not available for the apparent masses of the polypeptides.

2.6.3   Green plants: PSII resembles purple bacteria

The active centre of PSII in green plants (RCII) has recently been extracted from PSII (originally from spinach chloroplasts) by Nanba and Satoh (1987), and contains four polypeptides. Two of these carry a protohaem IX group between them and are identified as cytochrome b-559. The other two polypeptides, D_1 and D_2, had previously been shown to be analogous to the L and M polypeptides of PBRC by comparison of the sequences of their genes (Hearst and Sauer, 1984). The six pigment molecules, probably four chlorophyll a and two phaeophytin a, match the four Bchl a and Bph a molecules of the purple bacterial centre (PBRC), but there is no crystallographic proof yet that the structures are at all similar, and some spectroscopic evidence bearing on the orientation and spacing of the pigments argues against the hypothesis.

RCII exists in an environment very different from PBRC, in that the PSII particle is much larger, containing at least two other Chl a-containing polypeptides, CP47 and CP43 (disregarding CP29, described above as a linker protein, LHCIIα). These two Chl-proteins may be described as 'internal' or 'core antennas' of PSII. There is no Chl b, and the complex carries carotene (chiefly β-carotene), and no xanthophylls. PSII also contains small polypeptides (4.6–10 kDa) which are probably unpigmented. It is also associated in the intact membrane with the assembly known as the watersplitting apparatus (or oxygen-evolving system). This is a complex of unknown structure involving manganese (4Mn), calcium and chloride ions, protected by at least three water-soluble extrinsic polypeptides on the luminal side of the thylakoid membrane. Only the latter are coded by nuclear genes, the intrinsic polypeptides being all coded on the chloroplast genome.

2.6.4   Green plants: PSI—more chlorophyll attached
             to fewer proteins

RCI of green plants appears to contain two large Chl a-proteins, subunits Ia and Ib (83 and 82 kDa actual mass, see Table 2.2). These polypeptides carry, possibly, 90 Chl molecules, of which only three are known to take part in photochemistry. The bulk of the Chl forms a core antenna. Solvents, detergents and other treatments gradually remove Chl, leaving a photochemically active PSI particle with 40 Chl or fewer. In this way the Chl of PSI has been allocated to 'peripheral' and 'internal' regions. LHCI is attached to the RCI forming the PSI as normally encountered, with 200 Chl

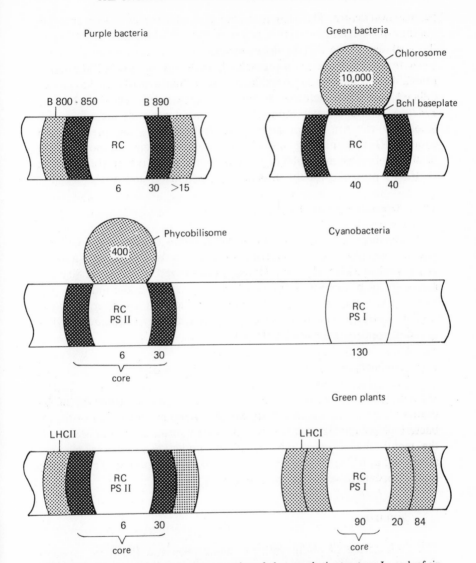

Figure 2.14. Diagram relating the common plan of photosynthetic structure. In each of six examples, the reaction centre, RC, (two polypeptides) is shown in white, with the number of chlorophylls indicated underneath. The RC may be associated with a core-antenna (dark stippling), or contain a large number of Chl/Bchl thus forming its own core-antenna. Light-harvesting (antenna) complexes (LHC, light stippling) may be intrinsic or extrinsic to the membrane. After Packham and Barber (1987).

per reaction centre. This structure may exist as a trimer as seen in some micrographs. Again, β-carotene is found. Subunits Ia and Ib of RCI span the thickness of the thylakoid membrane.

There are seven other, non-pigmented, small polypeptides in PSI ranging from 25 kDa (PAGE-mass) to 8 kDa. Allocating functions to these has proved difficult since the homologies are hard to trace between species (see Nelson, 1987, for a brief review). Subunit II has been tentatively identified with secondary electron acceptors (iron-sulphide clusters are present) and subunit III with binding sites for the soluble electron-donor protein plastocyanin. Two molecules of phylloquinone, and a further iron-sulphide centre are present, probably attached to subunits Ia and/or Ib.

2.6.5 *Green sulphur bacteria*

Less information is available about the reaction centres of the Chloro-biacae. The photoactive bacteriochlorophyll a is attached to a large polypeptide (65 kDa, Hurt and Hauska, 1984) together with some 80 other molecules of bacteriochlorophyll which may resemble the multiple pigments of PSI of green plants. The same identification is supported by the similar electron-transport intermediates (see Chapter 4), although phylloquinone appears to be absent (Hauska, 1988).

2.6.6 *Heliobacterium chlorum*

Pigment–protein complexes have not yet been isolated from the membranes of this new species, but it appears that it has an antenna consisting of bacteriochlorophyll g, in three distinct spectral forms at around 788 nm, in the membrane. There is also a minor unidentified antenna pigment absorbing at 670 nm and a single carotenoid, neurosporene. The reaction centre appears to resemble the PSI type.

2.7 Summary

The architecture of photosynthetic membranes and the three types of pigment complex (reaction-centre, core complex and antenna complex) are compared in Figure 2.14.

CHAPTER THREE

PRIMARY PHOTOPHYSICS
Times from 1 fs to 100 ps

3.1 Light absorption: formation of excited states of molecules

Molecules consist of atoms, held together by chemical bonds. A chemical bond depends upon electrons being shared in such a way that the (negatively charged) electron density is located between the (positively charged) atomic nuclei thus providing electrostatic attraction (a chemical bond). Electrons in atoms or molecules are not to be regarded as points, but distributed into waveforms known as orbitals. Orbitals have their characteristic size, energy and shape in space. Each orbital can hold two electrons, which must have opposite directions of their spin. Orbitals centred on one atomic nucleus are known as atomic or non-bonding orbitals, while those connecting two (or more) nuclei are molecular orbitals.

Light is quantised; that is, it is absorbed and emitted in packets of energy, quanta or photons. Monochromatic light may be regarded as a stream of photons of identical energies. The energy of the photon is given by:

$$E = h\nu = hc/\lambda$$

where h is Planck's constant, ν is the frequency, c the velocity of light and λ the wavelength.

When light is absorbed by photosynthetic pigments, the effect is to promote one electron from a low-lying energy level or orbital to a higher one. The primary event in photosynthesis is therefore the absorption of one quantum of light by a pigment molecule. This is an illustration of the Grotthuss–Draper principle: that light is only effective for photochemistry or photobiology when it is absorbed by a chromophoric group.

Since each orbital has its characteristic energy, a pigment only absorbs those wavelengths for which the energy of the photon corresponds to the difference between the energy levels of two orbitals. An absorption spectrum indicates the values of the energies of the available electronic transitions.

Colour, that is absorption in the 'visible' wavelength range, is relatively rare in organic materials, and is usually due to unpaired electrons or an extensive system of conjugated double bonds, the latter being important in photosynthetic pigments. The absorption at 1015 nm (near infrared) by the antenna of *Rps. viridis* represents probably the lowest-energy electronic transition in photobiology. In the opposite direction, in the ultraviolet below 400 nm, absorption becomes more general particularly in protein (aromatic amino acids at 260–290 nm, the peptide bond at 190–220 nm). Visible light therefore corresponds to a range in which electrons can be excited in specific materials without interference from general cell constituents.

The absorption spectrum of chlorophyll a (Figure 3.1a) shows two major peaks, corresponding to the first (670 nm) and second (430 nm) excited states of the π-orbital system. In each case the electron distribution is changed along the axis of rings I and III (Figure 2.8a) known as the Y-axis, and the major peaks are known as Q_Y transitions. The smaller peaks or shoulders on the blue (higher energy) side of the major peaks are due to Q_X transitions (the axis of rings II and IV), and vibrationally excited sub-states of both X and Y excitations. All the X-states and the higher Y-states rapidly lose energy as heat and are converted to the lowest energy major peak at the red-end of the absorption spectrum, that is around 678–690 nm for chlorophyll a, 770–890 nm for bacteriochlorophyll a (see Figure 3.1b) or up to 1015 nm for bacteriochlorophyll b.

The carotenoids (formulae, Figure 2.10; absorption spectra, Figure 3.2) show extensively conjugated double bonds and absorption maxima that to some extent occupy the 'window' between the blue and red absorption maxima of the chlorophylls. Their functions are discussed in section 2.32.

The thermal energy of molecules causes their atoms to vibrate about their mean positions. These vibrations typically have periods of 10^{-13}–10^{-12} s. The promotion of the electron to the higher orbital is much faster, completed in some 10^{-15} s. Promotion of the electron is more likely when the atoms are near the extrema of their vibration. These are the Franck–Condon principles. Because the electron distribution has enlarged, the excited bond is less attractive and the vibrations of the excited molecule are larger, that is, part of the energy of the absorbed light appears as thermal energy. The thermal energy is rapidly redistributed over all the bonds of the molecule. The broad appearance of visible absorption spectra in general is due to the spread of thermal energies, and the random influence of neighbouring atoms in solutions. Chlorophyll is unusually sharp; the breadth of the absorption curves of chloroplasts and chromatophores is

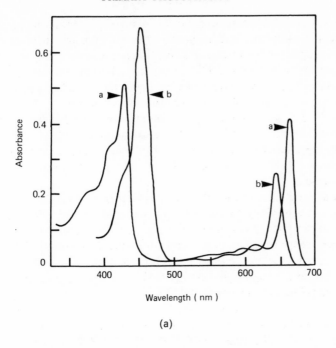

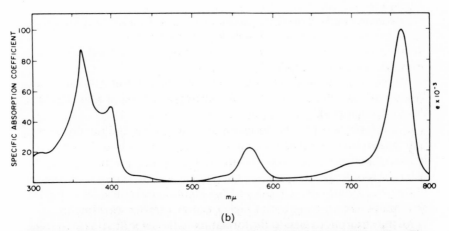

Figure 3.1. (a) Absorption spectra of solutions of chlorophylls a and b, in 80% acetone. The concentrations were $5.0 \mu g \, ml^{-1}$ and $7.9 \mu g \, ml^{-1}$ respectively; the pathlength was 1 cm. (b) Absorption spectra of bacteriochlorophyll a in ether, from Strain and Svec (1966). The ordinate is the specific absorption coefficient (calculated for a concentration of $1 \, g \, ml^{-1}$).

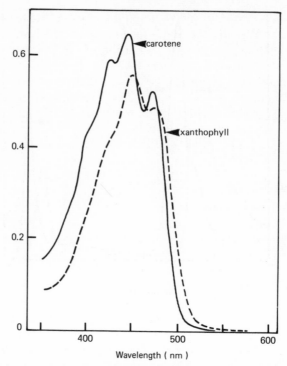

Figure 3.2. Absorption spectra of two carotenoid fractions from pea leaves, containing mainly β-carotene and lutein respectively. The concentrations were 2.27 μg ml^{-1} and 2.10 μg ml^{-1} respectively, and the pathlength was 1 cm.

due to the aggregated differences in the protein–chlorophyll complexes. Within each complex, the specific ordering imposed by the protein sharpens the absorption peak.

In chlorophyll, the second excited state is not stable and rapidly decays into the first excited state, losing all the excess energy as heat. This places a limit on the efficiency of a plant in capturing light energy from the sun: light of longer wavelength than say 700 nm is not absorbed, and only the energy of P680* or P700* (about 1.8 eV or 170 kJ mol^{-1}) is retained following the absorption of each quantum of higher energy (shorter wavelength).

All the absorptions lead to the formation of the same (first) excited state, within 10^{-15}–10^{-14} s. In this excited state there are now two orbitals containing one electron each. It is a 'singlet' excited state (a spectroscopic term) because the unpaired electrons retain their opposite spins. It is important that in normal photosynthesis the first singlet excited state is the operational one. It is possible, and is the normal event for photochemical reactions in solution, for a small proportion of singlet excited states to invert the spin of one of the unpaired electrons. This produces the triplet state, which has a much longer lifetime, long enough for other molecules (in solution) to approach and react with it. Such reactions, such as that with molecular oxygen, are deleterious in biological membranes and must be avoided: one role of the carotenoids is to collect and deactivate any chlorophyll triplet states that are formed, before they can react with O_2.

3.2 Possible fates of excited states

Excited states decay exponentially: the relationship:

$$Z_t = Z_0 \cdot \exp(-t/T)$$

describes the quantity Z decreasing as a function of time (t) and a characteristic lifetime T. That is, the rate of decay is proportional to the quantity present (a first-order process). The singlet state of chlorophyll (a single isolated molecule) has a natural lifetime of some 15 ns. Chlorophyll in condensed media has a much shorter lifetime, in which time it either loses its energy by means of (a) fluorescence, (b) transfer of the excitation energy to an adjacent chlorophyll molecule, (c) photochemisty in the reaction centre, or (d) a radiationless deactivation, producing heat. Each mode of decay generates another exponential term in the above equation. The transfer processes that we shall be considering have much shorter lifetimes, and measurement of the rate of decay of fluorescence reveals the rates of all the other decay processes currently operating.

3.2.1 *Fluorescence*

Fluorescence is the re-emission of a photon of light from an excited pigment molecule. We have seen that some of the energy of the original photon was lost as heat; a further loss occurs on fluorescence, because the electron returns to a vibrationally-high energy level of its ground state, according to the Franck–Condon principles. This is illustrated in Figure 3.3.

The result is that the fluorescent emission of the molecule has a maximum at a longer wavelength than the absorption maximum. For pigments such as chlorophyll at around 700 nm, the difference (known as Stokes' shift) is of

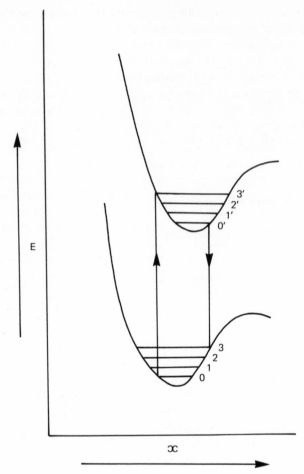

Figure 3.3. Absorption of a photon, formation of an excited state, and emission of a photon by fluorescence. The diagram represents graphically the energy of a chemical bond as a function of the distance between the two atomic nuclei. Two curves are shown, representing the ground state and first excited state. The vertical scale of the vibrational energy levels is exaggerated. The absorption shown is a 0–3′ transition; heat is lost as the vibrational energy decays, and the fluorescent emission is a 0′–3 transition. The fluorescent quantum has a lower energy and hence a longer wavelength (Stokes' shift).

the order of 7 to 10 nm. Often the fluorescent emission spectrum resembles a mirror image of the absorption spectrum.

Fluorescence is a drain on the supply of energy for photosynthesis, but it is in fact virtually negligible. For the investigator, however, fluorescence has been and remains a uniquely valuable means of studying the primary photophysics of the photosynthetic pigments.

3.2.2 Excitation energy transfer

Excitation energy can be transferred from one pigment to another. This is the basis of all antenna pigment action. For the strongly pigmented membranes in photosynthesis, the mechanism is that described by Th.W. Förster and known as *inductive resonance*. This depends on (a) the fluorescence spectrum of the excited (donor) molecule overlapping the absorption spectrum of the recipient, (b) the distance between the two molecules being of the order of 10 nm or less in chlorophyll (efficiency falls off with increasing distance as the sixth power) and (c) the orientation (transfer is most favoured when the directions of the absorption dipoles of the two pigments are parallel or anti-parallel).

We have seen (Chapter 2) that the antenna is an array of pigmented polypeptides packed together; their spacing and orientation are held by the protein framework, and the graded absorption maxima provide a path such that excitation energy is directed to the photoactive pigments in the reaction centres. Figure 3.4 shows the absorption spectra of the antenna and reaction-centre complexes of a purple bacterium; although the reaction-centre provides only a very small fraction of the total absorption of the cell, it nevertheless efficiently captures all the light energy absorbed. The close spacing, 1 to 2 nm, corresponding to a concentration of more than $0.1 \, mol \, dm^{-3}$, provides for efficient and rapid transfer. It is believed that transfers take some $10^{-12} \, s$, and therefore energy will reach the reaction centre within a few picoseconds, regardless of where the original photon was absorbed in the pigment bed.

3.2.3 Radiationless deactivation

A dilute solution of chlorophyll in say 80% acetone fluoresces visibly; about 30% of the photons absorbed are re-emitted, in a characteristic lifetime of some 5 ns. The remainder are converted to heat, by processes dependent on the conversion of excitation to vibrational energy, or by conversion of the excited state to the triplet state (in which the spin of the excited electron is inverted) or by quenching by interaction with oxygen molecules. By contrast, the deactivation rate in the photosynthetic membrance is very low, partly because the protein provides a controlled

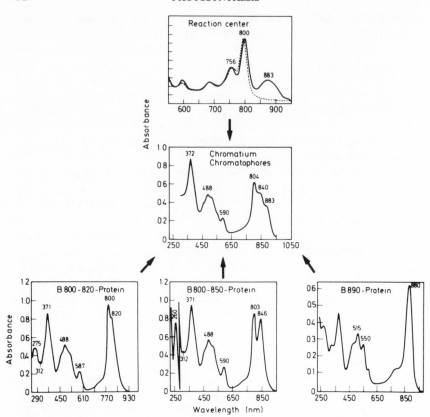

Figure 3.4. The pigment-protein composition of a purple bacterium, *Chromatium vinosum*. The absorption spectrum of the chromatophores (centre) is made up of the three antenna proteins (below) and the reaction centre (N.B. different wavelength scale) (above), which represents only a very small fraction of the total pigment. The dotted line shows the bleaching when the reaction centre is in the P^+ state. Note that all bands at wavelengths longer than 800 nm are due to bacteriochlorophyll a (compare Bchl a in ether, Figure 3.1b). From Thornber (1970).

environment, partly because the transfers are completed in some 100 ps, not allowing time for the slower process of deactivation to have effect.

3.3   Antenna chlorophylls in chloroplasts

Antenna complexes, for example the LHCII complexes from higher plants, when isolated, have a fluorescence at room temperature that is easily visible to the eye. In the natural state their fluorescence can only be readily detected with sensitive equipment. Fluorescence mainly comes from the PSII core complex (at ambient temperatures; at low temperatures the PSI core complex is the major fluorescent entity). Remember that the bulk of the chlorophyll in green-plant core complexes functions as an antenna. The chlorophyll in these complexes is differentiated, presumably by means of its local protein environment, into groups; the groups have differing but overlapping wavelengths for their maxima of absorption and fluorescence emission. The groups are resolved by calculating the best fit that can be obtained from a few standard curves, to the actual absorption spectrum of a photosynthetic membrane.

Where possible, experimenters separate PSI from PSII by means of non-detergent methods. Mechanical breakage can produce sealed, inside-out vesicles corresponding to grana membranes, with no PSI, and stromal thylakoid fragments enriched in PSI.

One approach to the relationships of the groups of core-chlorophylls is to measure the lifetimes of the fluorescence components due to each family of chlorophyll molecules. The sample is excited with a series of flashes from a laser, each flash having a duration of 1 to 30 ps. The fluorescence of the outermost group is expected to rise with the lifetime of the flash, and decay; as it decays the next group begins to fluoresce at a different wavelength, reaching a maximum from which it decays, followed by the third group, and so on. Because all the components overlap, and because the length of the exciting flash is comparable with the lifetimes of the early species, a computer-dependent process known as deconvolution is needed to resolve the data into components having consistent lifetimes and spectra. Results at present are subject to the limitation that it has to be assumed that fluorescence decay is exponential in form, and also, because of the statistical calculations, only three to five components can be reliably resolved at the present time.

The chlorophyll families are described by their maximum absorption wavelength, in nm (e.g. C674) or by the fluorescence maximum (e.g. F690):

LHCI: C674 (F690), C677 (F730)
CCI (core complex of PSI): C680 (F690) (80% of the total), C697 (F720) and C705 (F735)
LHCII: C670 (majority), C680 (F695)
CCII: C669 (33%), C676 (F685) (33%) and C683; phaeophytin Ph680–685.

3.3.1 *PSI*

The main emission from PSI at ambient temperature is at 690–695 nm, and has a short lifetime (12–20 ns). It is assigned to C680, quenched by rapid

54 PHOTOSYNTHESIS

transfer of its energy to other chlorophylls. It remains more or less constant as the temperature is lowered to 77 K, when the major emission is seen at 720 and 735 nm. The rise-times for F720 and F735 are equal to the decay of F690, and they are biphasic, the components being 12 ps and 75–100 ps. The data (numerical examples taken from Wittmershaus, 1987) have been interpreted on the basis that C680 passes its energy to C695 and C705 in parallel. These are the lowest energy chlorophylls in the system and hence trap the energy; thermal energy is needed to assist the uphill transfer to P700 (actually at 695 nm). At ambient temperature the transfer is rapid and the longer-wavelength fluorescence emission is effectively quenched, but at 77 K the barrier is sufficient for F720 and F735 to be detected. The slower component is due to excitation arriving from C680 in LHCI.

3.3.2 PSII

Preparations containing PSII show two (some reports show three) components of fluorescence decay at 681–685 nm. The longest-lived (2 ns) component shows the influence of chlorophyll b in its excitation spectrum and therefore comes from light absorbed by LHCII. The shorter form (1 ns) (and the third, 0.24 ns) are excited by chlorophyll a only. All these components are subject to fluorescence induction, that is, they increase in magnitude by up to six times when the PSII reaction centre is 'closed', that is, unable to accept the excitation energy because the electron transport has not completed the previous event. An emission has been detected at 695 nm, ascribed to the phaeophytin in the reaction centre (van Dorssen et al., 1987).

3.3.3 Excitation migration

If the PSII reaction centre is closed the excitation continues to migrate, at random, and is able to travel via the LHC to other reaction centres. This is particularly important in purple bacteria or between PSII centres in green plants. This is a concept known as the 'puddle' or 'lake' theory; the number of connected PSII cores is at least four (the puddle) but may increase greatly (the lake) when LHCII is rearranged in the membrane. To some extent excitation energy that arrives at a PSII centre but cannot be immediately used may be able to reach a PSI centre, in green plants (but see the discussion of 'lateral inhomogeneity', Chapter 6). This is the concept of 'spillover'. Spillover from PSI to PSII does not take place because first, the transfer is from pigments absorbing at 700 nm to others at 680 nm, an unfavourable direction (but not impossible), and secondly, the PSI reaction centre appears to catalyse a thermal deactivation when it is unable to use the energy for photochemistry. Energy that enters PSI complexes does not re-emerge. The propagation of excitation energy has been studied in crystals, and the term 'exciton' is commonly used for the quantum of excitation energy in a crystal. By analogy, it is often used for excitation energy in photosynthesis, but without physical precision, since the chlorophyll array is not crystalline.

3.3.4 Variables affecting the fluorescence yield

Both PSI and PSII fluorescence is increased when the reaction centre is closed by virtue of the primary electron acceptor being reduced, although the effect is much harder to demonstrate in PSI. A second factor is the electric field across the membrane, that is part of the energy store for ATP synthesis (see Chapter 6): an increasing field causes diminished fluorescence. Thirdly, LHCII can become phosphorylated and disconnected from PSII (the state I to state II transition, see Chapter 7) thus altering the balance of excitation between the photosystems. This may be an electrostatic effect, and it is observed that magnesium (Mg^{2+}) ions have an influence increasing fluorescence up to some 3 mmol dm^{-3}. Fourthly, the slowest effect (minutes) is brought about by photoinhibition, when CCII (particularly D1) is damaged by excess illumination. Fluorescence measurements provide a non-invasive probe of many photosynthetic features in widely differing time ranges.

3.4 Photochemical charge separation in reaction centres

In every case, the basic mechanism of reaction-centre photochemistry appears to be the same. There is a closely interacting pair of molecules of the primary pigment (Chl a, or Bchl a or b), known as a special pair, that becomes excited and passes an electron to an acceptor molecule. The pair becomes a radical cation. The cation loses its previous absorption spectrum, and a bleaching at that wavelength is seen; the pigment is said to be photoactive and is denoted by P: thus P700, the first to be discovered (by B. Kok in PSI of green plants), showed a bleaching at 700 nm. P also conveniently stands for 'pair'.

The acceptor is also a pigment and is in close proximity to P, held rigidly by the protein framework. The acceptor may be symbolised by I, and the photoactive pigment by P, so that the reaction is:

$$[P.I] \xrightarrow{\text{excitation}} [P^*.I] \rightarrow [P^+.I^-].$$

Both the cation P^+ and the anion I^- are free radicals.

Although the reaction centre of purple bacteria (P870 or P890 in forms containing Bchl a, P960 with Bchl b) was observed subsequently to P700, it was elucidated rapidly and now forms the basis of our thinking concerning reaction centres in general. The spectra of P^+ obtained by means of ESR and ENDOR (electronic nuclear double resonance) indicated that the unpaired electron was distributed over the π-orbital system of both members of a (hypothetical) special pair; the existence of the pair was confirmed by X-ray crystallography.

The finding of the close analogy between the PBRC and PSII makes it virtually certain that P680 is a special pair of chlorophyll a molecules.

P700 presents more difficulty. Crystals have only recently been obtained, and the assembly is much larger than any protein successfully analysed so far. Early ESR and ENDOR data suggesting delocalisation of the unpaired electron in $P700^+$ was challenged, and re-interpreted to show the reverse, that is, a single Chl^+ component. However, circular dichroism studies show a split exciton effect expected of a (dissymmetric) interacting pair. Studies by means of ADMR (absorption-detected magnetic resonance) showed that the triplet-minus-ground-state absorption spectrum resembled a dimer rather than a monomer of chlorophyll.

There has been much interest in minor derivatives of chlorophyll a, isolated from PSI. Chl-RCI is chlorinated on a methene bridge, and modified in ring V, and was isolated in proportions of 1:1 to P700 by Dörnemann and Senger (1985). An epimer of chlorophyll at C-10 (in ring V) known as Chl a' was isolated in proportions of 2:1 by Watanabe et al. (1987). The chlorination of Chl-RCI has been admitted to be a preparation artefact, leaving the possibility still viable that P700 is based on a dimer of Chl a', or Chl a in an environment where enolisation occurs in ring V.

All the P forms known are single-electron donors, and apart from P680 have redox potentials around 0.45 V at pH 7, equilibrating with artificial oxidising agents such as ferricyanide. P680 has a redox potential of about 1.1 V at pH 7.

The nature of the immediate electron acceptors I was inferred from ultra-rapid spectroscopy of isolated purple bacterial reaction centres (see Parson, 1987 for a review). The technique involves 'pumping' a sample with a picosecond flash, and then 'probing' with a white, picosecond flash after a preset delay of tens of picoseconds. The probe flash after passing through the sample is recorded on a vidicon surface by a spectrograph. Since the primary photochemical reaction takes place very fast, it was necessary to ensure that subsequent acceptors (quinones) were fully reduced so that the electron would stop at I^-. The simplicity of the structure (see Table 2.5

and Figure 3.4) allowed the observation of each of the pigments without mutual interference, or time delay in an antenna. There are only six pigment molecules, four Bchl a or b, and two bacteriophaeophytin (Bph).

The bleaching of P870 and the appearance of the new maxima due to P870$^+$ were found to occur at the same speed as the loss of Bph and the formation of Bph$^-$. No faster changes could be seen, although the electronic structure of P870 is complex and the changes may depend for their great speed on interactions with other Bchl molecules or protein.

The precise arrangement of the six pigment molecules (Figure 3.6) was determined by means of X-ray crystallography (references in Chapter 2). The special pair of closely associated Bchl molecules can be easily distinguished, and the two Bph molecules. The remaining two Bchl molecules are positioned between the P-pair and the Bph; they are regarded

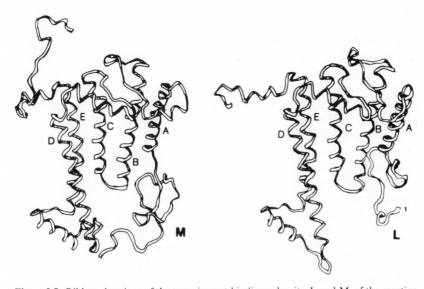

Figure 3.5. Ribbon drawings of the two pigment-binding subunits, L and M, of the reaction centre of the purple bacterium *Rps viridis*. The ribbon indicates the course of the peptide backbone, omitting side-chains. The molecules are seen from the edge of the membrane; the E-space (periplasm) is at the top and the cytosol (P-space) below. There are five membrane-spanning α-helical sections, labelled A–E in each unit. A sixth helix exists between D and E on the cytosolic side. The N-termini are both on the cytosolic side. The histidine residues (L-173, H-200) that each bind one of the two Bchl of P960 are located at the top (periplasmic end) of helix D in each case, and the L-unit is to be imagined placed against unit M forming a sandwich with the pigments (see Figure 3.6) inside. The figure may be compared in content and style with the representation of PSII in Figure 4.6. From Parson (1987) based on Deisenhofer *et al.* (1985).

as in some way forming a conduction path for the electron without giving rise to any recognisable reduced (anionic) form. They also, of course, have the same antenna capability as any other pigment in the system.

The Bchl molecules of the purple bacterial reaction centre are complexed to two proteins, L and M (Figure 3.5). It appears that although there appear to be two pathways for the electron from P* to the two Bph:

$$P^* \rightarrow Bchl_L \rightarrow Bph_L$$

$$(P^* \rightarrow Bchl_M \rightarrow Bph_M)$$

in fact only the path carried by the L protein is active.

Very recently, a complex has been isolated from green plants that appears to be the reaction centre of PSII. It contains about four Chl a molecules, two Ph a and one β-carotene. It is photoactive. The principal proteins, known as D_1 and D_2 (referring to a 'diffuse' appearance), are clearly related in their amino-acid sequence to the L and M proteins of purple bacteria, and there is great attraction in supposing that the analogy holds further in that P680 is formed by a special pair of chlorophyll a molecules and that the phaeophytin is the immediate electron acceptor. Figure 4.6 shows a model predicted on this basis.

In PSI of chloroplasts, it has not been possible to obtain preparations containing so few pigment molecules. However, good crystals of trimeric PSI from a cyanobacterium have been recently obtained (I. Witt et al., 1988) and may allow a structural X-ray analysis, although the unit size of 600 kDa is unprecedentedly large.

The earliest spectral change is observed at 690 nm, synchronously with the exciting light pulse, and decays with a lifetime of 15 ps, synchronously with the appearance of $P700^+$. The 690 nm bleaching is ascribed to the formation of the excited state of chlorophyll in the inner antenna. Some of the excited antenna forms the triplet state which lasts for 1–2 ns (Giorgi et al., 1987). The early electron acceptors (A_0, A_1) have short lifetimes, and their spectra can usually be observed only when subsequent acceptors ($A_2 = F_x$, iron-sulphur centres, see the next chapter) are experimentally disabled or reduced. The immediate acceptor from P700 (I in the universal symbolism, A_0 in this case) has been argued to be a Chl a molecule absorbing at 670 nm (Mansfield and Evans, 1985), or alternatively one absorbing at 693 nm, found by Nuijs et al. (1987) to be formed from $P700^+$ in 32 ps.

The same problem is found in the green sulphur bacteria, where the acceptor from P840 (the special pair of Bchl a molecules) may be Bchl a. In

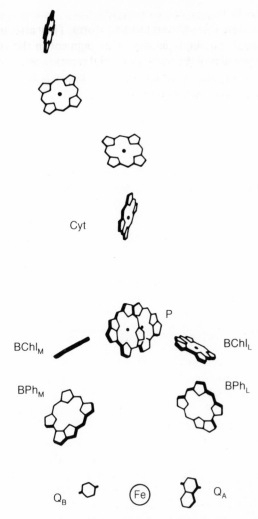

Cyt

BChl$_M$ P   BChl$_L$

BPh$_M$ BPh$_L$

Q$_B$ Fe   Q$_A$

Figure 3.6. Perspective diagram of the arrangement of cytochrome, pigment and quinone molecules in the purple bacterial reaction centre. Based on the X-ray crystallographic solution of the reaction centre of *Rps viridis* obtained by Deisenhofer *et al.* (1985). The four polypeptide chains. (L, M, H and cytochrome) are not shown, for the sake of clarity. The side chains of the tetrapyrrole rings are also omitted; the edges are shown by a thicker line. The four haem groups at the top (the E surface of the membrane) belong to the extrinsic C-cytochrome. They are the electron donors to P960 (labelled P). P is made up of two Bchl b molecules, in close overlapping juxtaposition. The route of the electron transfer is to the right, via Bchl$_L$, BPh$_L$, Q$_A$ (menaquinone), Fe and Q$_B$ (ubiquinone). UQ diffuses and exchanges with Q$_B$ in the plane of the membrane. From Parson (1987), taken from the stereographs in Deisenhofer *et al.* (1985).

Heliobacterium chlorum the acceptor may be Bchl c (not Bchl g) (Nuijs *et al.*, 1985), unexpectedly since Bchl c, like Chl b in green plants, was thought to be exclusively an antenna pigment.

With the formation of the ion-pair $[P^+.I^-]$, the primary photophysical period is over. Up to this point no molecular movement has been required, and all changes have been purely electronic.

CHAPTER FOUR

ELECTRON TRANSFER WITHIN REACTION-CENTRE COMPLEXES
Times from 4 ps to 0.15 ms

The electron is a negatively charged particle and migrates according to the local electrostatic potential, being more likely to move so as to occupy an orbital of the same or a neighbouring molecule with a lower energy, that is, with a more positive potential. In so doing, the electrical work done appears as heat. The greater the heat released at any stage, the less reversible the transfer. The path followed by an electron in escaping from a reaction centre may be studied by means of electrochemistry, a key principle of which is the concept of redox potentials.

4.1 Redox potentials

The electric potential at a point is the quantity of work required to bring a unit charge from infinity to that point. An electron (negatively charged) moves spontaneously from a position of high negative potential to one of lower potential, and the energy difference is released as heat. *Reducing agents* have a tendency to lose electrons in chemical reactions, and have a negative potential with respect to *oxidising agents*, which gain electrons.

Many reduction–oxidation (redox) reactions can be divided into two half-reactions, which describe the oxidant and the reductant respectively. It is possible to arrange the half-reactions in a series, with the most oxidising at the top and the most reducing at the bottom (most negative) end. This is the electrochemical series. Each half-reaction (describing a particular reagent under particular conditions) can be allotted a numerical value of redox potential (see Table 4.1). The standard hydrogen electrode (shown in Figure 4.1) has a defined potential of zero, and all other half-reactions are related to it by means of an apparatus of the kind shown in the Figure 4.1.

Table 4.1 Standard redox potentials of substances of biological interest. The redox potential listed for each couple is defined, if not actually measured, with respect to the standard hydrogen electrode at pH 0, for example, as shown in Figure 4.1. Thus, couples with positive potentials are oxidising with respect to those with negative potentials.

Couple	pH	E^{o}_{pH} volts	
P680: $Chl \rightarrow Chl^+ + e^-$		1.17	
$Z \rightarrow Z^+ + e^-$	in situ	> 1.0	
$2\,H_2O \rightarrow O_2 + 4H^{2-} + 4e^-$	5.0	0.9	
$2\,H_2O \rightarrow O_2 + 4H^+ + 4e^-$	7.0	0.81	
P700: $Chl \rightarrow Chl^+ + e^-$		0.49	
P870: $Bchl \rightarrow Bchl^+ + e^-$		0.4 to 0.5	(in purple bacteria)
$K_4Fe\,(CN)_6 \rightarrow K_3Fe\,(CN)_6 + e^- + K^+$		0.44	
$Fe^{II/III}$ between Q_A and Q_B		0.40	
Plastocyanin: $Cu^{I/II}$		0.37	
Cytochrome b-559 (high potl. form)		0.38	
Cytochrome c-555		0.35	(in *Chromatium vinosum*)
Cytochrome f: $Fe^{II/III}$		0.3	
Cytochromes c, c_2		0.3	(in purple bacteria)
Cytochrome c_1		0.22	
Rieske FeS centre		0.29	
UQ or PQ: $QH_2 \rightarrow Q + 2H^+ + 2e^-$		0.10	
Cytochrome c-553		0.01	(in *Chromatium vinosum*)
Cytochrome b-559 (low potl. form)		0.0	
UQ or PQ: $Q_B + 2e^- + 2H^+ \rightarrow Q_BH_2$	in situ	0.0	
$H_2 \rightarrow 2H^+ + 2e^-$	0.0	0.000	
Cytochrome b_6	7.0	−0.05 and −0.15	
PQ: $Q_A + e^- \rightarrow Q_A^{\cdot -}$	8.6	−0.13	
$H_2S \rightarrow 2H^+ + 2e^-$	7.0	−0.27	
$NAD(P)^+ + 2e^- + 2H^+ \rightarrow NAD(P)H$	7.0	−0.32	
$H_2 \rightarrow 2H^+ + 2e^-$	7.0	−0.41	
Ferredoxin		−0.42	
FeS centre: F_A		−0.54	
$\quad\quad\quad\quad F_B$		−0.5	
Phaeo $+ e^- \rightarrow$ Phaeo$^-$		−0.6	
P680: $Chl^* \rightarrow chl^- + e^-$		−0.65	
Bphaeo $+ e^- \rightarrow$ Bphaeo$^-$		−0.7	or below (in purple bacteria)
FeS centre: F_X		−0.705	
A_1: phylloquinone	in situ	−0.9	
A_0: $Chl + e^- \rightarrow chl^-$		< −1.0	
P700: $Chl^* \rightarrow chl^+ + e^-$		−1.28	

Many of these values are subject to some discussion, or show inter-specific variation. Chloroplast components are discussed by Mathis and Rutherford (1987), and bacterial components (as indicated) by Dutton (1986).

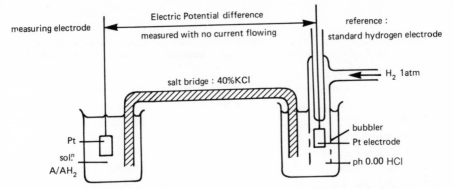

Figure 4.1. Principle of determination of redox potential. The measuring electrode consists of a platinum surface in contact with a solution containing oxidised and reduced forms of a couple, which will react at the surface, in a solution of pH 0 (or a stated pH). The reference electrode shown is a standard hydrogen electrode (defined as zero potential). If the electrolytes are not identical then some such *salt bridge* as shown is required to control any junction potential. The redox potential of the couple A/AH_2 is determined for a range of relative concentrations (redox titration), progressing in both directions. The potential obtained when A and AH_2 are at standard concentrations is the standard potential at the pH and temperature of the solution.

Thus, if a typical half reaction is:

$$e^- + X \rightleftharpoons X^-$$

then the electromotive force of the cell in which one half is the standard hydrogen electrode and the other half is made up of a redox couple X and X^- (at unit concentration of each) is equivalent to the standard redox potential of X. The usual symbol is E^o or E_m, and the unit is the volt. If the reagent X forms the acid HX:

$$e^- + H^+ + X \rightleftharpoons HX$$

then the standard conditions must include unit concentration of H^+ ions, that is, pH 0.

In biological circumstances pH 0 is unrealistic, and by convention pH 7 or other stated pH is used and the symbol for the standard potential is differentiated by a 'prime' thus $E^{o'}$ or $E_7^{o'}$.

In the example of the couple X/X^- above, one electron takes part. In general, n electrons may take part in a redox half-reaction. This affects the variation in the potential as the relative proportions of X and X^- are changed. For any mixture of X and X^-, the potential is given by the equation

$$E = E° + (RT/nF)\ln([X]/[X^-])$$

where square brackets stand for the concentration, and F is Faraday's constant of $96\,460\,C\,mol^{-1}$ (equal to the charge of Avogadro's number of electrons). It is important that the concentration ratio is expressed as the oxidised form of the couple (X in the example) divided by the reduced form (X^-). The equation follows from the law of mass-action, and to generalise further, if the equation of the half-reaction involves other than one molecule of the oxidised or reduced form, the numbers of molecules involved appear as indexes to the concentration terms. Thus for the hydrogen electrode:

$$0.5\,H_2 \rightleftharpoons H^+ + e^-$$

$$E = 0.00 + (RT/F)\ln([H^+]/[H_2^{0.5}])$$

which shows that as the concentration of molecular hydrogen is increased, the potential falls, varying as the square root. Note that when dealing with gases the standard state may be chosen either as unit molar concentration in solution or 1 atmosphere gas pressure ($101.3\,kPa$), but different values will apply for the standard redox potential. The hydrogen electrode is only defined as zero for the standard state of gas at $101.3\,kPa$ pressure.

In the example where H^+ ions are involved, and if $m\,H^+$ occur per electron in the equation of the half-reaction, then the dependence on pH is given by:

$$E_{pH}^{o\prime} = E° - 2.303m(RT/F)\,pH$$

where 2.303 is the conversion factor from natural to base 10 logarithms, and the negative sign arises from the definition of pH as the negative logarithm of the hydrogen ion concentration. Qualitatively, the equation means that couples of the X/XH type have a more negative standard redox potential at higher pH values. Thus, for the standard hydrogen electrode where m is unity, the redox potential falls from zero (defined) at pH 0 to $-0.42\,V$ at pH 7. This is still true regardless of the reaction mechanism: whether or not hydrogen reacts directly with the compound X, or whether X gains an electron first, and acquires the proton from the solution subsequently.

A knowledge of redox potentials allows us to predict or rationalise the behaviour of electrons in moving between the various sites in the photosynthetic membrane.

A newly discovered redox material can be characterised by redox titration, that is, the determination of the potential obtained when the oxidised/reduced ratio is varied, at different pH values. Provided that the material behaves reversibly, meaning that the same values are obtained irrespectively of the direction of the titration, this establishes the values of $E^{o'}$, m and n. These enable an estimate to be made of the chemical nature of the electron-carrying site, and of the role of the material in the biological process. For example, cytochromes have characteristic spectra in the oxidised and reduced states, and because the only change is the gain or loss of one electron from the iron atom, there is usually no change of potential with pH. Nevertheless, changes in the redox state of a prosthetic group such as haem can influence the dissociation constant of groups in the protein, resulting in pH-dependent potentials found for example with cytochrome b.

4.1.1   Cytochromes

Cytochromes are proteins which take part in electron transport reactions and possess a form of haem as a prosthetic group. They were first observed by McMunn in 1886 by means of a simple spectroscope; in the reduced state cytochromes have sharp absorption bands in the visible spectrum. They were named by D. Keilin in 1925 who assigned them to the classes a, b and c on the basis of the wavelength of the α-band. Thus cytochrome c had an α-band at 550 nm, cytochrome b at 563 nm and cytochrome a at 608 nm. Each of these types spawned a family as related cytochromes were found elsewhere. Apart from the wavelength of the α-band, the types can be

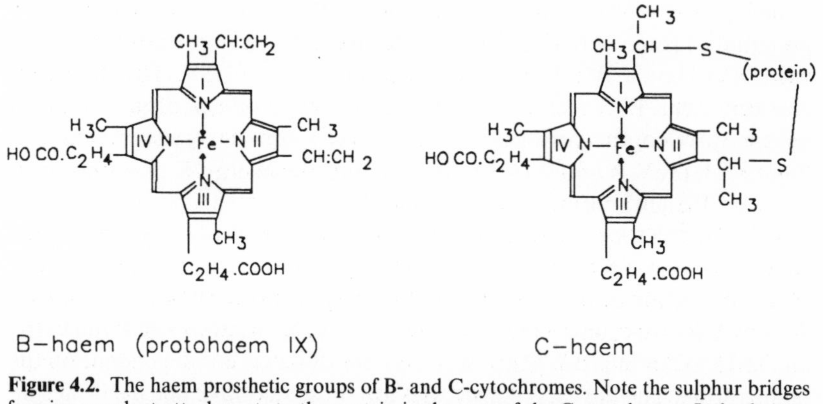

B—haem (protohaem IX)   C—haem

Figure 4.2. The haem prosthetic groups of B- and C-cytochromes. Note the sulphur bridges forming covalent attachments to the protein in the case of the C-cytochrome. In both cases the Fe of the haem is coordinated to two histidine-N atoms above and below the plane of each haem.

distinguished chemically; B-type cytochromes have protohaem IX, the same as haemoglobin, coordinated from the iron atom to the nitrogen of histidine, but whereas in haemaoglobin there is only one such link, leaving a coordination position open for oxygen, in B- and C-cytochromes there are two and no other ligand can bind (Figure 4.2a). In C-type cytochrome there are two additional, covalent bonds from the sulphur of cysteine groups to the vinyl side-chains of the haem (Figure 4.2b). Hence, haem can be dissociated from B-cytochromes by means of acid-acetone, but not from C-cytochromes. The haem of A-cytochromes is more variable, often having isoprenoid groups attached. A-type cytochromes are not represented in the photosynthetic membrane, although as cytochrome oxidase particles they coexist with photosynthetic complexes in purple non-sulphur bacteria. B- and C-type cytochromes, in photosynthetic membranes, do not bind substrates (and are therefore not regarded as enzymes) and act only by exchanging one electron, alternating between oxidation states Fe(II) and Fe(III) of the central iron atom.

Cytochromes have been named in several ways, either by suffixing a serial number (c, c_1, c_2, etc.) or by the wavelength of a convenient α- or γ-band wavelength (b-559, etc.). Cytochrome f in chloroplasts is a C-type cytochrome (cytochrome c_6).

4.1.2   Chlorophyll

The essence of light-driven electron transport is that chlorophyll, normally neither a notable oxidising nor a reducing agent, becomes a powerful reducing agent when excited. The couple P*/P$^+$ in P680 has a redox potential of some -0.6 V. On the other hand P$^+$ is a powerful oxidising agent: the couple P$^+$/P has a redox potential of $+1.2$ V. The difference between them, 1.8 V, represents the 1.8 eV energy of a photon of light of 680 nm wavelength, corresponding to the zero–zero energy represented in Figure 3.3. (1 eV is equal to $1.6\,10^{-19}$ J, and the energy E of a photon is given by Planck's formula $E = h\nu = hc/\lambda$).

In Chapter 3 the primary reaction of photosynthesis was described as the transfer of one electron from a chlorophyll to an acceptor (I), which in different reaction centres could be a phaeophytin molecule or a chlorophyll. This transfer was completed in 4 ps in the purple bacterial reaction centre, and in 15–32 ps in green plant photosystem II; it was not dependent on the temperature. The next set of processes are those that take place in times up to 0.15 ms. They are in general temperature-dependent and are influenced by fluctuations in the form or conformations of the polypeptide chains in the reaction-centre complexes. Nevertheless, reactions in this timescale do not depend on molecular diffusion, and take place entirely within the multi-subunit particle, the reaction centre, in all the examples known. We are concerned here with chains of electron donors and acceptors existing in the reaction-centre complexes. By their means, electrons are conducted away from the $[P^+.I^-]$ ion pair to (in the case of purple bacteria and PSII) diffusible quinones, or (in the case of PSI and green sulphur bacteria) the diffusible protein ferredoxin. Each type of reaction centre has its chain of acceptors.

In contrast, not all reaction centres contain a chain of electron donors. PSI of green plants, and some purple bacteria, do not appear to have any intrinsic donors of electrons to P^+ but react directly with diffusible proteins such as plastocyanin or cytochrome c.

4.2 Quinones—the electron acceptors for the reaction centres of PSII in green and purple bacteria

The purple bacteria, and almost certainly PSII of green plants, have a pair of quinone molecules attached to the L, M and D_1, D_2 proteins respectively.

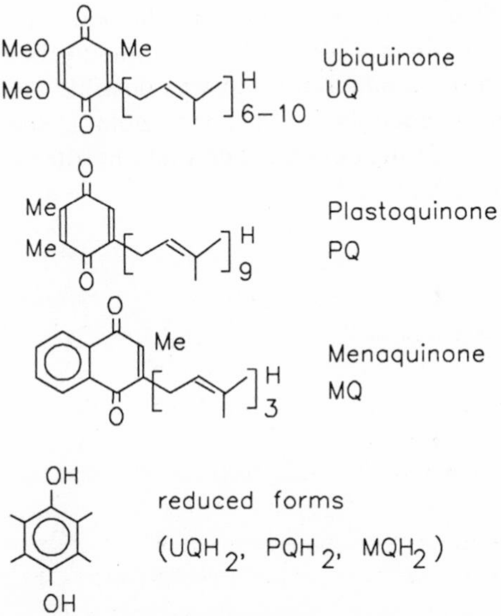

Figure 4.3. Diffusible quinones in electron transport.

These quinones are located on the side of the membrane nearer to the cytosol of the bacterial cell or the stroma of the chloroplast. Chloroplasts employ plastoquinone, bacteria ubiquinone or menaquinone (formulae, Figure 4.3). The first quinone to receive electrons from Bph or Ph is known as Q_A. It is relatively fixed in position (to the M-subunit in the purple bacteria), and is only able to accept one electron to form the semiquinone radical-anion (Figure 4.4). This was shown by comparing the spectrum of the intermediate with authentic plastosemiquinone formed by pulse radiolysis. The transfer to Q_A takes some 200 ps (Holten et al., 1978). Presumably the protein to which it is attached prevents it from becoming further reduced or taking up H^+ ions to form a quinol. The second quinone molecule, Q_B, is able to accept two electrons in sequence from Q_A, and acquires two protons to form the corresponding quinol. The electronic part of the transfer takes place in 0.15 ms (Parson, 1969). The quinol then exchanges with an oxidised quinone from the diffusible pool in the lipid phase of the membrane. The H^+ ions are taken from the cytosol/stroma phase. Q_B can be said to act as a two-electron gate.

There is an atom of iron situated between the quinones Q_A and Q_B. It was at first assumed that this transition element was assisting the electron transfer between the quinones, but experimental removal of the iron or replacing it with other metal atoms does not inactivate the reaction centre.

The first crystal structures elucidated for the reaction centres of purple bacteria (Rps. viridis) did not contain Q_B; presumably, being exchangeable, it has been lost in the crystallisation. It was possible to force Q_B into the crystal and hence to identify the site. It is notable that in Rps. viridis, although the bulk, diffusing quinone is ubiquinone, the tightly bound Q_A is

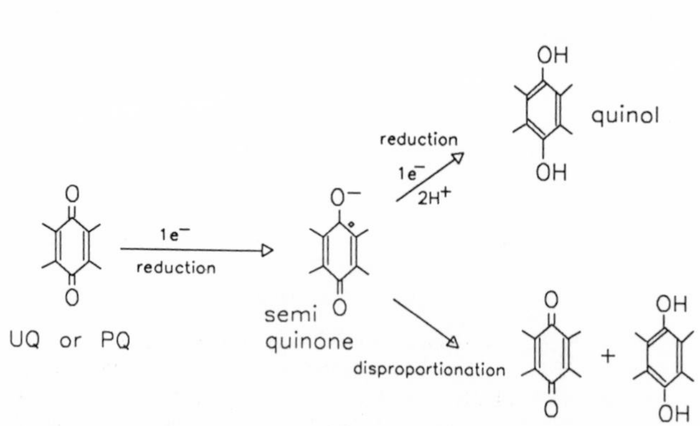

Figure 4.4. The formation of semiquinone.

menaquinone. Both Q_A and Q_B are present in the crystal structure of *Rb. spheroides.*

There is no crystal structure yet for PSII of green plants. The four-peptide preparation available at present has lost both Q_A and Q_B. The models at present in circulation are based on the sequences of amino acids in the D_1 and D_2 proteins, deduced from the genes that code for them in the chloroplast genome, and calculations of the likely three-dimensional structure based on analogy with known examples.

4.2.1 *Fluorescence induction*

The symbol Q originally denoted a quencher of fluoresence. In green plant PSII, when white light is turned on, fluorescence begins at a relatively low level (level O) and rises in about 1 s to a maximum (level P). This change in fluorescence yield (the Kautsky effect, Figure 4.5) is known as fluorescence induction and is interpreted as being due to the action of a 'quencher' which

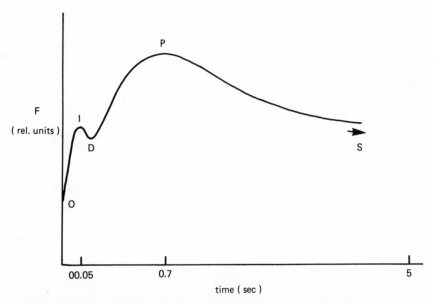

Figure 4.5. Fluorescence induction (the Kautsky effect). The curve is drawn on a non-linear time-scale to illustrate features observed at millisecond and second times. F, fluorescence intensity; D, dip; I, intermediate peak; O, zero-time level of F; P, peak and S, steady value. Further changes, at 0.5 min (M) and 1.5 min (T) occur in leaves and are not here. (See Lavorel and Etienne, 1977, pp. 253–4.)

is initially in the oxidised state but becomes reduced by photochemical action. The reduced quencher does not quench, and the fluorescence increases. The quencher is therefore an early electron acceptor in PSII. The induction is greatly speeded up in the presence of the herbicide diuron (DCMU, dichlorophenyldimethylurea) which displaces Q_B from the D_1 protein so that Q_A cannot be oxidised by Q_B. Instead of the pool of plastoquinone becoming reduced over a period of 1 s, photochemical action stops and fluorescence induction is completed in milliseconds with the reduction of one molecule of Q_A per reaction centre.

The artificial electron acceptor silicomolybdate is able to act in the presence of diuron, and modifies the protein in some way rendering the Hill reaction with ferricyanide immune to diuron. The normal protein structure is believed to be stabilised by HCO_3^- ions in the Q_A-region (quite apart from the importance of CO_2 in photosynthetic metabolism); if bicarbonate is removed by washing chloroplasts with formate solutions, activity declines sharply.

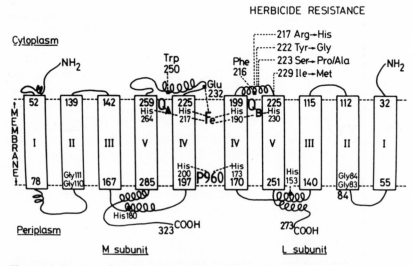

Figure 4.6. Structure of the D_1 and D_2 proteins of the reaction centre of PSII. The binding sites of chlorophyll a (as P680) (histidines D_1-198, D_2-198), plastoquinone and Fe are based on the close analogy with respect to amino acid sequence with the purple bacterial L and M proteins. The five membrane-spanning α-helices deduced from the sequence are numbered I to V and correspond to A–E in Figure 3.4. The numbers are the amino acids counting from the N-termini. Also marked are amino acids which are the sites of binding of some herbicides, and mutations which confer herbicide-resistance (see Figure 4.7). From Barber (1987) based on Trebst (1986).

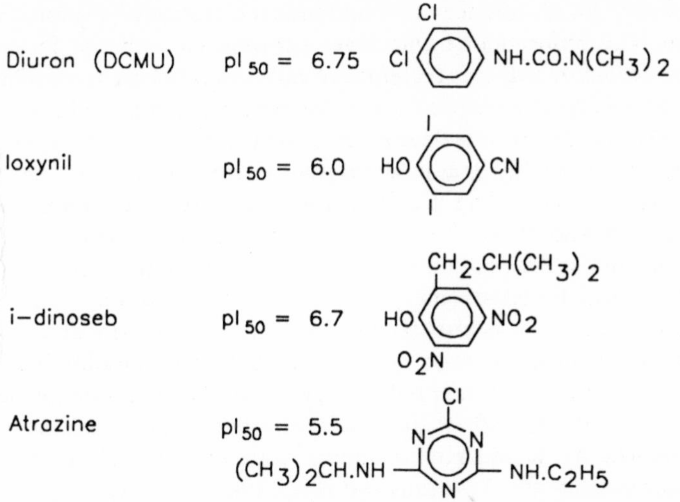

Diuron (DCMU) $pI_{50} = 6.75$

Ioxynil $pI_{50} = 6.0$

i–dinoseb $pI_{50} = 6.7$

Atrazine $pI_{50} = 5.5$

Figure 4.7. Structures of some PSII herbicides. Two families are represented: those containing the —C—N-group (atrazine and diuron), and phenol derivatives. They bind to the D_1 protein (Figure 4.6) as non-identical sites which mutually interfere, and prevent electron transfer from Q_A to Q_B, probably by displacing Q_B. See Trebst (1988).

Besides diuron, other groups of herbicides have been shown to bind to the D-proteins in PSII (see Figures 4.6, 4.7). Analysis of herbicide behaviour is assisted by noting that part of the D_1 protein can be removed by digestion with the proteinase trypsin, without damaging or inactivating the photosynthetic capability of the thylakoid membrane. The action of trypsin is prevented if certain herbicides are applied first, and the action of herbicides such as diuron is prevented if trypsin digestion is carried out first.

4.3   Ferredoxins—the electron acceptors in the reaction centres of green plant PSI and green sulphur bacteria

4.3.1   *Electron transport within PSI*

Electron transport within the PSI complex commences with the formation of the ion-pair $[P700^+.A_0^-]$, in which the electron acceptor A_0 is probably a third molecule of Chl a (Chapter 3). The redox potential of $P700/P700^+$ is $+0.49$ V (Sétif and Mathis, 1980); that of Chl/Chl^- is lower than -1.0 V (*in vitro*, that of A_0 is not known). The formation of the ion-pair therefore accounts for at least 1.5 eV of electrical work out of the 1.8 eV available from one photon at 700 nm; it is probably much closer since the ion-pair can back-react to re-excite the antenna chlorophyll.

The reduction of the second acceptor, A_1, is observed when the subsequent acceptors are chemically reduced or disabled by anionic detergent. Nuijs *et al.* (1987) measured a decay time of 32 ps for the primary acceptor, presumably the time for the reaction A_0 to A_1. When the absorption spectrum obtained following a flash of light is subtracted from the 'dark' spectrum, a difference spectrum is obtained in the ultraviolet showing a maximum at 290 nm (an aromatic material, not a tetrapyrrole). A characteristic EPR (electron paramagnetic resonance, or ESR electron spin resonance) spectrum is also obtained which is narrowed if the sample is treated with deuterated water (the material contains replaceable hydrogen). Two molecules of the naphthoquinone phylloquinone (vitamin K_1) are present (Figure 4.8), and are extractable in ether. Ether extraction appears to inactivate A_1. Hence phylloquinone is an attractive but by no means certain candidate for A_1 (Mansfield *et al.*, 1987). It is expected to have a redox potential of approximately -0.9 V; the back-reaction with P700$^+$ is still significant (time constant of 120 μs).

Three redox centres then follow, which have been detected and characterised by their EPR spectra at low temperatures. All three are iron–sulphur centres (bound ferredoxins) with very low potentials, labelled X, A and B (or F_X, F_A and F_B); A has the least negative redox potential (-0.54 V) and is preferentially reduced by a single-turnover flash. If A is already reduced, B (-0.59 V) is reduced. The above values are for spinach and vary in other plants. The reduction time for A and B is less than 1 ms at room temperature. The centre X (-0.705 V) has been identified with entities observed by means of optical absorption-spectroscopy and known as A_2, or P430 (bleaching at 430 nm), reduced in less than 100 ps. Alternatively, P430 could correspond to centres A and B. The kinetic assignments are still disputed (see the review by Mathis and Rutherford, 1987).

The A and B centres are carried by one or both of two small polypeptides

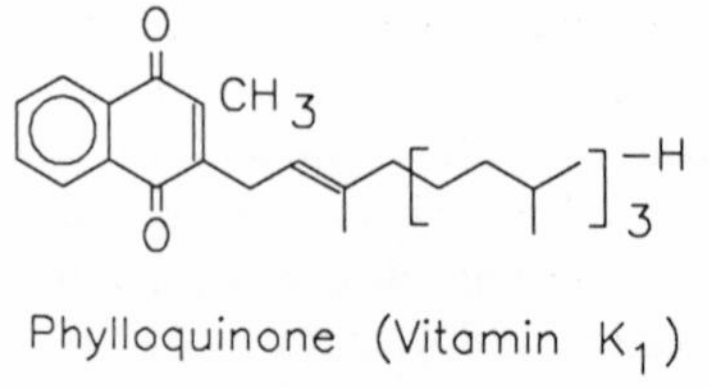

Phylloquinone (Vitamin K_1)

Figure 4.8. Structure of phylloquinone, vitamin K_1, proposed as the second acceptor A_1 in photosystem I.

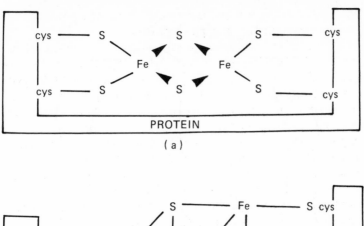

(a)

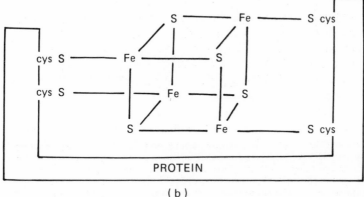

(b)

Figure 4.9. Iron–sulphur centres. (a) Fe_2S_2 centres in green-plant type ferredoxin, and in the Rieske centre. The cysteine residues are part of the protein. (b) Fe_4S_4 centres of the PSI acceptors F_X, F_A and (possibly) F_B, and of ferredoxins from green and purple sulphur bacteria.

that are intrinsic to the membrane, but do not span it (subunit II or III). A_0, A_1, and X, as well as P700, are carried by one or both of the large polypeptides, subunits Ia and Ib. All three iron–sulphur centres are probably of the Fe_4S_4 type (Figure 4.9b) which accounts for the analysis of 12Fe, 12S in PSI. Centre X may have a different structure from that of A and B on the basis of Mössbauer spectra. The present state of speculation is represented in Figure 4.10.

4.3.2 Ferredoxin is the diffusible electron acceptor

PSI reduces a diffusible protein that exists in the stroma of the chloroplast, ferredoxin. Ferredoxin is a single polypeptide chain of about 12 500 kDa, that

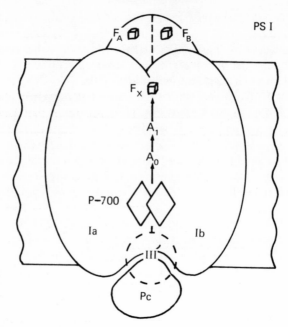

Figure 4.10 Diagram showing a possible arrangement of electron-transport intermediates and polypeptides in PSI. It is controversial in respect of the number of copies of subunits Ia and Ib, the dimeric nature of P700, the small polypeptides that carry centres F_A and F_B, and the role of subunit III in binding plastocyanin (PC). The luminal (E) space is below, the stromal phase above the plane of the membrane. After Mathis and Rutherford (1987).

carries an iron–sulphur cluster of the Fe_2S_2-type (Figure 4.9b). Its redox potential, -0.4 V, is about equal to that of the hydrogen electrode at pH 7, and it is likely that all the low-potential iron–sulphur clusters have a reduced form in which all the iron is in the ferrous Fe(II) form, and an oxidised form which may be regarded formally as containing one Fe(III) atom. Ferredoxins, when isolated, lose their prosthetic group in acid solution giving hydrogen sulphide and ferrous ions. They have characteristic red colour and visible absorption spectrum (with relatively low extinction coefficients), but they are observable by means of EPR only at liquid helium temperatures.

Under conditions of iron-deficiency, some algae are able to synthesise *flavodoxin*, a flavoprotein, which takes over the functions of ferredoxin.

4.3.3   The RC of green sulphur bacteria resembles PSI of green plants

The Chlorobiaceae have proved harder to study than other organisms. It can be said that the active pigment, bacteriochlorophyll a in P840, is associated with a pair of large polypeptides (of the order of 65 kDa) and the photochemistry involves acceptors that have similar absorption spectra to those of A_0 and A_1 in green plant PSI. These acceptors may therefore be a bacteriochlorophyll and a naphthoquinone (Hauska, 1988, points out that phylloquinone is absent from *Chlorobium*, but a hydroxylated naphthoquinone would have been harder to detect). It is not certain whether FeS groups are part of the reaction centre (only a Rieske-type centre has been reported so far), but ferredoxin (the bacterial type, Fe_4S_4) is certainly the diffusible acceptor.

The reaction centre of *Heliobacterium chlorum* (P798) has not been extensively studied so far, but it appears that an early acceptor may be bacteriochlorophyll c (reduced in 200 ps), followed in 500 ps by a more distant acceptor that may be an FeS centre suggesting a resemblance to PSI.

4.4   Electron donors

4.4.1   Four families of C-cytochromes

Cytochrome c itself was first observed in mitochondria where it is soluble, located in the space between the envelope membranes. It is also found in Gram-negative bacteria in the periplasmic space (between the cell membrane and the cell wall). A similar form is known as cytochrome c_2. Either kind may be found in purple bacteria (including the green *Chloroflexus*).

Cytochrome c_1 is bound to the bc complex in bacteria and mitochondria, covered in Chapter 5. The corresponding form in chloroplasts is cytochrome f, properly classified as c_6.

Cytochromes c-552 and c-555 each account for two haem groups attached to the same polypeptide chain. The polypeptide is attached on the periplasmic side of the reaction centre of the purple bacteria *Rps. viridis* and *C. vinosum* (see Table 2.5). Other purple bacteria do not have this component.

Cytochrome c-553 is known in cyanobacteria, and some green algae. It is soluble in the thylakoid lumina and acts in the same way as plastocyanin. Although similar in function to the c, c_2 group, they cannot be interchanged. The green sulphur bacteria, and also *H. chlorum*, also have a cytochrome *c*-553 acting as an electron donor to their reaction centres, but it is membrane-bound. (See the review by Dutton, 1986.)

4.4.2   *Purple bacteria*

The crystal structure of the reaction centre of *Rps. viridis* shows a large polypeptide chain attached on the E-side (periplasmic) of the reaction centre, covering the pigments. This polypeptide carried four haem groups, and is identified as cytochromes *c*-552 and *c*-555. It does not appear to be anchored in the membrane as the L, M and H polypeptides are, and so is an extrinsic component, but it is tightly bound nonetheless. The four haem groups are in a line directed towards the P960 pair of Bchl b molecules, and provide a pathway for electrons to reach the oxidised $P960^+$, in a time of 270 ns (Holten *et al.*, 1978). This cytochrome is absent in *Rhodobacter sphaeroides*, and there is no bound protein or other electron donor to the P890 pair of Bchl a molecules.

In both the above species, and in purple bacteria generally, the reaction centre is reduced by means of cytochrome *c* ($E'_0 = 0.254$ V) or c_2 ($E'_0 = 0.3$ V), which are small (13 kDa) proteins existing in solution in the periplasmic space, outside the cell membrane but within the cell wall.

4.4.3   *Water is the ultimate electron donor to PSII in green plants*

The oxidising side of PSII is still rather in an 'alphabetic' state, in which entities are known from spectroscopic or other studies, but not character-ised chemically; they are labelled with alphabetic letters that do not necessarily agree between different laboratories.

The immediate donor to P^+ is known from its electron paramagnetic resonance (EPR) spectrum which is known as signal II, in several forms, formed or detected at certain speeds. It appears likely from electron–nuclear double resonance (ENDOR) studies that signal II is produced by two entities, labelled D and Z, that have been tentatively identified as tyrosine residues at positions 160 and 161 on the D_2 and D_1 polypeptides respectively. The oxidised form is tyr^+ (Figure 4.11). It has been suggested that D is not on the main electron transfer path, but acts as a reservoir.

The oxidation of water to molecular oxygen (O_2) provides the electrons that reduce oxidised Z and D. The process is known to involve manganese, from the work of Pirson and of Kessler which showed that manganese-deficient algae (*Ankistrodesmus*) lost their power to generate oxygen, and

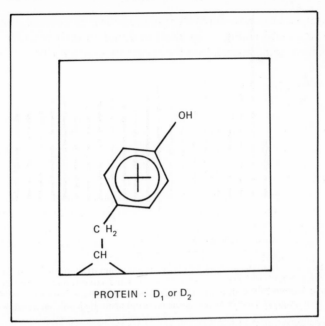

Figure 4.11. Model for a tyrosine residue to act as Z, the electron donor to photosystem II. The tyrosine is expected to be that shown in Figure 4.6, at position 160 in the D_1 protein. The same structure may account for the associated centre D, on the D_2 protein.

regained it on addition of manganese chloride to the culture, in less time than would have been needed to synthesise protein. Chloroplasts treated with 0.8 M tris (base) solution lose their manganese, and their oxygen-evolving competence, reversibly to some extent. Loss of manganese does not inactivate PSII, since alternative electron donors can be provided, such as hydrazine, hydroxylamine (NH_2OH) or diphenylcarbazide, and PSII still carries out photochemistry, reducing acceptors such as benzoquinone or the dye 2, 6-dichlorophenol-indophenol.

The reaction centre drives electrons one at a time. The oxidation of water to molecular oxygen has to be achieved four electrons at a time to avoid dangerous intermediates such as peroxyl radicals. This necessitates a four-electron gate. Such a gate was observed in experiments in which dark-adapted chloroplasts were subjected to very short, saturating flashes of light, sufficient to turn over each reaction centre once and only once. The oxygen resulting from each flash was registered as a pulse in a wide-area oxygen-electrode. The pattern of pulses was a damped oscillating train with

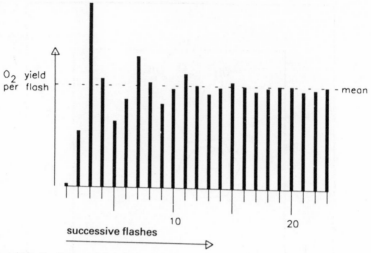

Figure 4.12. Oscillation of the yields of oxygen from a train of short saturating flashes. The bars represent the quantity of oxygen registered on a large-surface oxygen electrode after the nth flash, from cells of the green alga *Chlorella*. Flashes were at 300 ms intervals, and followed a dark pre-incubation of 3 min duration. A clear four-fold periodicity is seen. From Joliot *et al.* (1969).

period 4 and maxima on the 3rd, 7th, 11th, etc., flashes (see Figure 4.12). This was interpreted as a four-electron store S passing through the states S_0, S_1, S_2, S_3, and S_4, the state S_4 immediately decomposing to yield S_0 and O_2. Each transition from S_0 to S_1, etc. required one turnover of the reaction centre, that is, it was a one-electron reaction. Hence the periodicity of 4. It was further proposed that states S_0 and S_1 were equally stable in the dark, but that S_2 and S_3 decayed, so that dark-adapted material contained three times as much S_1 as S_0. Hence. the maxima on the third flash (Joliot *et al.*, 1969; Kok *et al.*, 1970).

Since manganese is a transition metal capable of taking several oxidation states, it was considered likely that the store S was related to the manganese complex in PSII. An additional EPR signal can be observed at very low temperatures, with up to 19 component lines, which is consistent with a multinuclear manganese complex containing higher oxidation states. It accompanies S_2 in flash series.

Manganese can also be studied with X-ray techniques EXAFS (extended X-ray absorption fine structure) and absorption-edge. These have not yet led to precise models, but are consistent with oxo-bridged bi- or tetra-nuclear clusters.

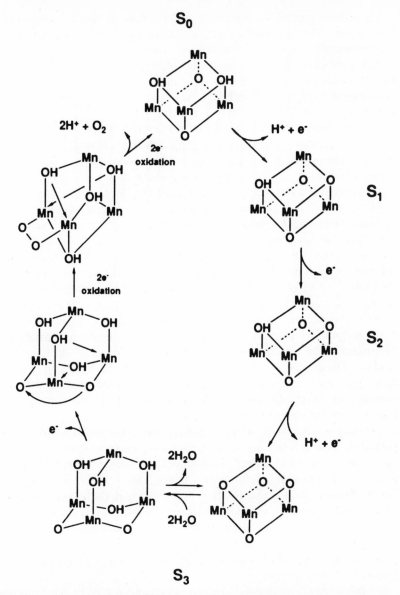

Figure 4.13. The oxygen evolving system; model of Brudvig and de Paula (1987). The structure shows a hydrated tetramanganese oxide with a cubane-structure. Progressive oxidation $(S_1, S_2, S_3$ and $S_4)$ produces oxygen and regenerates the low oxidation-level (S_0). Note the need for a change in conformation of the cluster.

Theoretical models for the S-system have been produced from several laboratories. They involve two or four manganese atoms. There are in fact four manganese atoms per P680, but they need not all be involved in the S-system. One of the considerations is the timing of the release of the four H^+ ions from the two molecules of water. These protons are released into the intraluminal space of the thylakoid membrane. They do not follow a simple pattern of one proton per flash $(1, 1, 1, 1)$, but rather $1, 0, 1, 2$. This may not reflect the working of the S-system, however, since it is possible that dissociable groups on the proteins may provide temporary sites for protons. An example of one model (Brudvig and de Paula, 1987) for a possible tetranuclear manganese complex is shown in Figure 4.13.

Studies have compared the sequences of the D_1, D_2 proteins with the L, M proteins of bacteria (which of course do not have an oxygen-evolving system). Amino-acid residues such as histidines have been found which could provide attachment for a tetranuclear manganese cluster such as the one illustrated. The cluster is believed to be protected by three water-soluble polypeptides (PAGE-masses 17, 23 and 33 kDa) that project into the luminal space of the thylakoid, probably as a cap over the manganese cluster. The cap in turn is stabilised by Ca^{2+} and Cl^- ions. It is possible to break thylakoids with ultrasound or the French pressure cell, in such a way that a proportion re-seal themselves inside out. The inside-out fraction can be isolated, and the three (or perhaps four) cap proteins detached and studied. These proteins are often referred to as water-oxidising proteins (their genes are labelled *wox*), but the identification was premature. It would appear that the D_1-D_2 pair of proteins carries all the prosthetic groups required for the electron transport process.

This discussion has ignored the cytochrome b-559 which is an intimate part of the PSII reaction centre. There is no satisfactory hypothesis for any action on the main line of electron transport, and indeed the redox potential (it has two forms with potentials of $+0.08$ V and $+0.38$ V) rules out such action. There is no analogue for cytochrome b-559 in the bacterial reaction centre, and it may be needed in chloropasts for deactivating the high-potential intermediates, if electron transport is restricted.

4.4.4   *Plastocyanin is the electron donor to PSI in green plants*

The oxidised reaction centre, $P700^+$, is reduced by a protein, plastocyanin, that is soluble in the aqueous intraluminal phase. There is no intrinsic electron donor forming part of PSI. In most cyanobacteria and some algae, such as *Chlamydomonas mundana*, cytochrome c-553 (also known

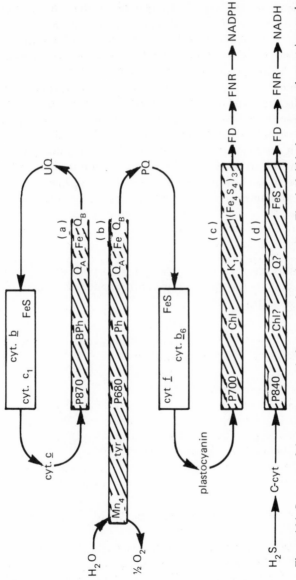

Figure 4.14. Summary of electron transport in four types of reaction centre. The shaded block represents the protein-complex containing the reaction centre. (a) Purple bacteria, (b) and (c) photosystems II and I of green plants and (d) green bacteria. The reactions to be described in Chapter 5 are indicated for the sake of context, omitting the non-cyclic process in (a) and the cyclic process in (d).

as cytochrome c-552) is found instead of plastocyanin, and in other species, such as *C. reinhardtii*, when they are stressed by conditions of copper deficiency, cytochrome c-553 is produced interchangeably with plastocyanin. This is analogous to the ferredoxin-flavodoxin alternation (4.3.1).

Plastocyanin is a globular molecule of 10 500 Da, carrying one copper atom towards one end. The amino-acid sequence is known, as is the tertiary structure from X-ray crystallography. The surface of the protein (which depends on the nature of the locally-exposed amino-acid residues) varies; there is an acidic patch and a hydrophobic patch, but it has been difficult to demonstrate the method of coupling with PSI, if indeed a lasting association is formed. A small membrane polypeptide on the luminal side of PSI has been suggested to act as a plastocyanin docking point. The reaction of plastocyanin with oxidised P700 has two distinct rate constants, which may be interpreted as showing the difference between properly-docked as opposed to randomly-passing plastocyanin.

4.5 Summary

Figure 4.14 compares the internal electron transport, and the electron acceptors and donors, for the reaction centres under discussion.

CHAPTER FIVE

ELECTRON TRANSPORT BY DIFFUSIBLE MOLECULES
Times from 1 ms to 20 ms

5.1 The ubiquitous cytochrome *bc* complex: the quinol cytochrome *c* reductase

In green plants, the slowest part of photosynthesis was detected by an experiment performed by Emerson and Arnold in 1932 in which leaves or cells were illuminated by a succession of very short flashes. It was found that the maximum rate of oxygen production or carbon dioxide uptake could be obtained with repetition rates of 100 per second, even though, with a flash duration of $10\,\mu s$, the plant was only receiving light for $1/1000$ of the time. The slowest step was (much later) traced to the reduction of a C-type cytochrome, cytochrome *f*, by the plastoquinol formed by PSII. Cytochromes are relatively easy to observe since their sharp α-bands in the reduced state can be measured spectroscopically in the window between the blue- and red-absorbing peaks of the chlorophyll. Cytochrome *f* in the reduced state has its α-band at 554.5 nm.

Plastoquinol does not reduce cytochrome *f* directly. The cytochrome is contained in a complex which is entirely distinct from the reaction-centre complexes, and which does not contain chlorophyll. The complex also contains cytochrome b_6 (two haems on one 23 kDa polypeptide, together with a 17 kDa polypeptide which together appear to be a homologue of mitochondrial cytochrome *b*) and a polypeptide carrying an Fe_2S_2 centre known as the Rieske centre. The Rieske centre differs from ferredoxin in that its potential is some 0.7 V higher (0.29 V, pH-independent between pH 6.5 and 8.0), and this is probably due to its containing two Fe(III) atoms in its oxidised state, and (formally) one Fe(III) and one Fe(II) in the reduced state. The Rieske centre is directly reduced by the plastoquinol, and subsequently reduces the cytochrome *f*. The reduced cytochrome *f* then reduces the (soluble) plastocyanin, which diffuses in the intraluminal

84 PHOTOSYNTHESIS

aqueous phase to PSI. The *bc*-complex is acting as an oxido-reductase enzyme, plastoquinol: plastocyanin reductase.

The polypeptide chains of cytochromes f (one) and b_6 (two), and the Rieske protein, span the membrane, and Figure 5.1 shows their possible arrangement. Note that the Rieske protein and cytochrome f have only one membrane-spanning section each, and their bulk, where the electron-

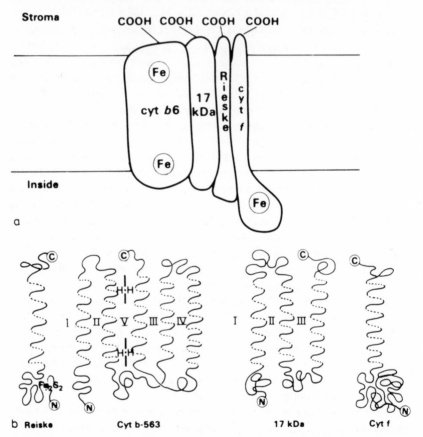

Figure 5.1. Diagram of the possible structure of the cytochrome $b_6 f$ complex. The lower part shows the membrane-spanning α-helices of the four proteins drawn in the upper part. The topography was determined by means of proteinases and antibody labelling, and the location of the helices from gene sequences and hydropathy-index plots. Note that the two haems of cytochrome b_6 are towards opposite sides of the membrane (they are coplanar and normal to the membrane plane), and that the Fe_2S_2 group of the Rieske protein and the haem of cytochrome f are in the aqueous environment in the thylakoid lumen. Bacterial *bc* complexes are similar except that the cytochrome *b* includes a section corresponding to the chloroplast 17 kDa unit as its *C*-terminus From Anderson (1987).

transport groups are located, lies in the thylakoid lumen space. Cytochrome b_6 on the other hand, is almost entirely located within the thickness of the membrane, with five spanning helices which carry the two haem groups, and a 17 kDa polypeptide, with three helices, that almost certainly represents a fragment detached during evolution. The complex also carries a protein kinase specific for LHCII, and forms a loose extrinsic complex with the ferredoxin-NADP reductase. These associations may be significant.

The purple bacteria contain a very similar complex, which is reduced by ubiquinol. A Rieske centre is present, which reduces in turn the bound cytochrome c_1 and the soluble, diffusible, cytochrome c. The reduced cytochrome c diffuses in the periplasmic space back to the oxidised reaction centre where it reduces $P870^+$ either directly or via the bound cytochrome c-555.

These complexes are clearly analogous to the complex III of mitochondria, which is in effect identical in its operation and composition to the complex of purple bacteria, being a ubiquinol-cytochrome c oxidoreductase. Furthermore, purple non-sulphur bacteria are facultative aerobes, and when air is present they lose their photosynthetic reaction centres and develop cytochrome oxidase particles, and NADH-ubiquinone reductase particles, and behave in many respects as mitochondria. It appears safe to generalise and to refer to a common type of (cytochrome) b, c complex in chloroplasts, purple and cyanobacteria, and mitochondria. Such comparisons (see Barber, 1984) are important in establishing evolutionary relationships (see Chapter 1).

There is insufficient evidence to be sure of a similar complex operating in the green sulphur bacteria, but a Rieske centre has been detected together with cytochromes c-550.5 and b-562, which have redox potentials comparable with cytochromes c_1 and b respectively in mitochondria making it a reasonable hypothesis. The diffusible quinone in the membranes of the green bacteria is menaquinone, and the hypothetical bc complex would be a menaquinol-cytochrome c-553 reductase.

Two points should be noted about the operation of bc complexes. First, the quinol carries 2H, but only gives up electrons to the complex. The H^+ ions are released on the periplasmic side of the bacterial membrane or the luminal side of the thylakoid membrane. This is part of the mechanism for the production of ATP. The second point is that the number of protons released is greater than the number of electrons transported, in purple bacteria and mitochondria, and the complex is operating in a more complicated way than that outlined above.

The quinol (UQH_2 or PQH_2) carries two electrons and two protons, but gives them up one at a time. This means that after the first electron has been given up, the quinol is in the free-radical 'semiquinone' state (Figure 4.4). The semiquinone is much more powerfully reducing than the quinol. It is able to give up its second electron, not to the Rieske centre (which is still reduced from the first electron transfer) but to one of the two haem groups of the B-cytochrome (in the region of -0.05 V). The electron is conducted through the protein complex, to the other haem group (approximately 0 V) of the B-cytochrome. This brings it to the membrane face on the cytoplasmic (matrix, stroma) side of the membrane where there is a bound quinone molecule. The quinone becomes reduced in two stages, picking up protons from the aqueous phase. The resulting quinol becomes unbound (by exchange with a quinone molecule) and diffuses across the membrane to the first-mentioned quinol binding site, reacting as before. One (oxidised) quinone molecule finds its way back to the reaction centre (or mitochondrial complex I or II) thus completing the quinone diffusion cycle in the thickness of the membrane, and causing, altogether, four protons to be translocated from the cytoplasmic to the periplasmic aqueous phases. The process is known as the Q-cycle (Mitchell, 1976, see Figure 5.2). The account given above is not proved in detail, and alternative schemes can be

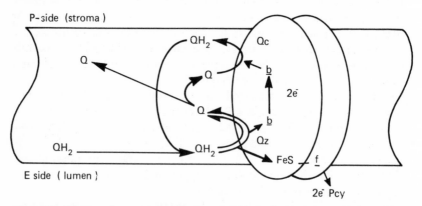

Figure 5.2. Diagram of one variant of the Q-cycle. Ubi- or plastoquinol is oxidised at the Q_z site on the E-side of the membrane, in two stages, via the semiquinone (not shown). Two electrons are released, one to the Rieske centre and one to the B-haem; two H^+ are released to the medium. The quinone resulting from the oxidation is reduced at the Q_c site at the P-side of the membrane, taking up $2H^+$ and $2e^-$ from the b-haem. The quinol is reoxidised at the Q_z site, completing the balance as shown, and the quinone returns to the reaction centre complex. The object of proposing Q-cycles is to explain (a) the high H^+/e^- ratios associated with transport though bc complexes, (b) the anomalous kinetics of cytochrome b, and (c) the existence of semiquinones. The Q-cycle may however not function in the steady state of chloroplast electron transport.

drawn. The evidence is that stable semiquinones can be detected, and that experimental one-electron oxidation of the bound C-cytochrome in bacteria and mitochondria leads to a detectable reduction of the B-cytochrome (provided the reoxidation of the B-cytochrome is blocked by the antibiotic inhibitor antimycin A).

It is striking that in models of the cytochrome b protein, the two haems are bound between the same two membrane-spanning helices, in such a way that the haem groups are coplanar and span the membrane. There is an aromatic amino-acid residue between them that could assist electron transfer from one haem to the other.

In chloroplasts there are some differences of detail. For example, no stable semiquinones have been detected, only two protons appear to be translocated under steady conditions per plastoquinone molecule, and antimycin A does not inhibit the (non-cyclic) electron transport. There is room to doubt the operation of the Q-cycle in chloroplasts, and it may depend on the overall redox balance of the thylakoid. The aromatic amino acid between the b-haem groups is missing.

There are many compounds available that inhibit electron transport in bc complexes. Many of these can be seen to act on either the quinol-oxidising site (where the Rieske centre is reduced) or on the quinone-reducing site (part of the Q-cycle). Thus DBMIB (Figure 5.3) is in the former category, while antimycin A is in the latter. However antimycin A is not effective in the b_6f complex; 2-heptyl-4-hydroxyquinoline N-oxide (HQNO) acts in all cases, and inhibits the reoxidation of the B-cytochrome.

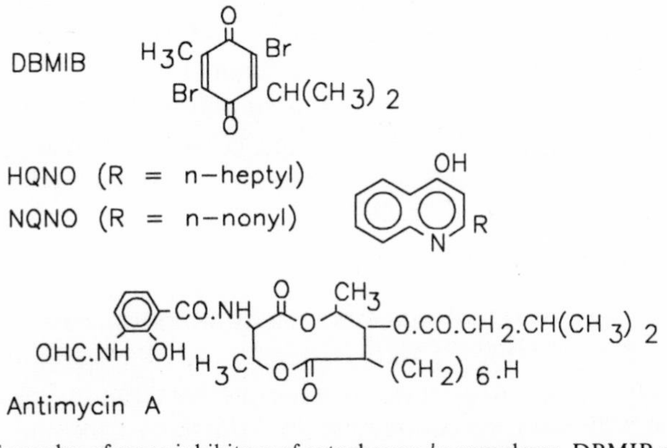

Figure 5.3. Formulae of some inhibitors of cytochrome bc complexes. DBMIB competes at the Q_z site (see Figure 5.2); HQNO and antimycin A inhibit the Q_c site (but antimycin A does not inhibit the b_6f complex in chloroplasts).

5.2   Patterns of electron transport: cyclic or non-cyclic

5.2.1   Purple bacteria

Cyclic. The purple bacteria have a simple photosynthetic system of a reaction centre and the bc complex, connected by ubiquinone that diffuses between the two in the membrane, and cytochrome c that diffuses in the periplasmic space. This provides for a closed loop as shown in Figure 5.4a. The energy for the electron circulation comes from the light absorbed by the reaction centre. The only effect of such a cyclic system is that the bc particle causes H^+ ions to be pumped from the cytoplasm to the periplasmic space, two pairs of H^+ ions per pair of electrons transported. One pair of H^+ results from the oxidation of UQH_2 to UQ, and the other pair from the Q-cycle. The H^+ ion gradient that is set up is a store of energy that can be used for ATP production, or for other purposes such as active transport of amino acids into the cell by exchange with H^+ or the operation of flagellar motors.

Non-cyclic. Some purple bacteria can couple the oxidation of environmental reductants such as sulphide or hydrogen to the electron-transport chain. These compounds bring about the reduction of cytochrome c, and via the reaction centre, the reduction of ubiquinone to ubiquinol. The membrane contains, in addition to the complexes already described, another complex which is closely analogous to complex I of mitochondria. It catalyses the reduction of ubiquinol by the reduced coenzyme NADH, and H^+ ions are pumped into the periplasmic space at the same time, by a mechanism that is not well understood. However, the photosynthetic system, in the light, maintains the H^+ gradient at its maximum, and keeps the ubiquinol pool reduced. Under these conditions the NADH-ubiqinone reductase runs backwards, driven by the H^+ ion gradient energy, and NAD^+ is reduced to NADH. In this way the environmental reductant is oxidised and the coenzyme reduced (Figure 5.4b). This arrangement is known as reverse electron transport.

Respiration. The purple non-sulphur bacteria (Rhodospirillaceae) are able to convert from anaerobic photosynthesis to aerobic respiration, that is from a phototrophic to a chemotrophic lifestyle (Figure 5.4c). The cell induces the formation of a complex which oxidises cytochrome c by means of molecular oxygen. Synthesis of photosynthetic pigments stops, and the cells become almost colourless after several divisions. Several types of

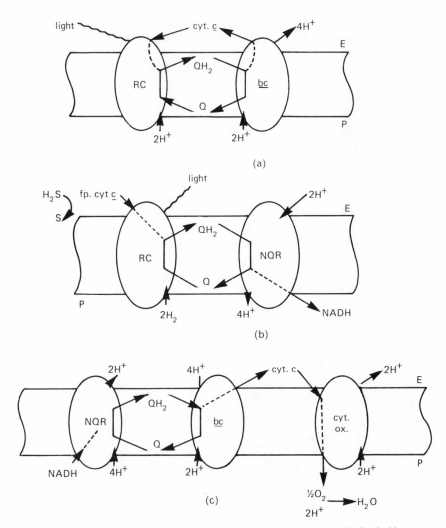

Figure 5.4. Overall photosynthetic electron transport in purple bacteria. (a) Cyclic. (b) Non-cyclic, in which hydrogen sulphide is oxidised by a flavocytochrome c, which reduces the reaction centre. NAD is reduced to NADH by the complex NADH:ubiquinone reductase (NQR), working against its normal direction. The energy is mediated by the hydrogen ions (see section 6.6). (c) Respiratory, showing NQR oxidising NADH, with the eventual reduction of oxygen to water by cytochrome oxidase. Note that the bc complex is not directly involved in (b), nor the photosynthetic reaction centre in (c). The number of hydrogen ions transported by each complex is controversial (see Figure 6.1).

cytochrome oxidase complex occur in the eubacteria, including the
cytochrome a–a_3 type found in mitochondria in eukaryotic cells, and the
electron-transport chain is driven by the energy of the overall process:

$$NADH + H^+ + 1/2\,O_2 \xrightarrow{\text{membrane}} NAD^+ + H_2O.$$

H^+ ions are pumped out of the bacterial cell, or the mitochondrial matrix,
by the same ubiquinol-dependent process as in photosynthetic electron
transport, but in addition both complexes I and IV pump protons, resulting
in $ATP/2e^-$ ratios of the order of 3.

5.2.2 Chloroplasts

Chloroplasts (including always cyanobacterial cells) have both types of
electron-transport process. The *non-cyclic* type generates oxygen and
reduces ferredoxin. PSI transfers electrons from water to plastoquinone;
plastoquinone diffuses to the *bf* complex and reduces the diffusible
plastocyanin. Reduced plastocyanin donates electrons to PSI, which
reduces ferredoxin (Figure 5.5). The coenzyme NADP is reduced by the

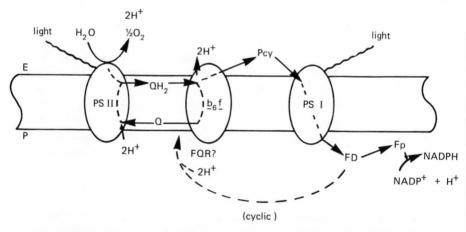

Figure 5.5. Overall photosynthetic electron transport in chloroplasts of green plants. PSII is
shown on the left acting as a water:plastoquinone oxidoreductase; electrons travel via the $b_6 f$
complex to PSI on the right, which acts as a plastocyanin:ferredoxin (FD) oxidoreductase.
Reduced ferredoxin reduces $NADP^+$ to NADPH by means of the membrane-bound
ferredoxin:NADP oxidoreductase (Fp). The cyclic pathway, involving only photosystem I and
the $b_6 f$ complex, via the hypothetical ferredoxin-plastoquinone reductase (FQR) is shown by
a dashed line.

reduced ferredoxin by means of ferredoxin-NADP reductase (FNR). The enzyme is somewhat loosely bound to the thylakoid membrane on the stromal side, probably by means of a 10 kDa polypeptide 'connectein' and a 17.5 kDa polypeptide anchored in the P-leaflet of the membrane. Pschorn *et al.* (1988) in a review suggest that the 17.5 kDa protein changes its shape in response to the rising proton gradient (see Chapter 6) and affects the binding of $NADP^+$ and ferredoxin to FNR; several seconds are required for activation. FNR contains flavin adenine dinucleotide as a prosthetic group. H^+ ion-gradient energy is not required for the reduction; the redox potential of ferredoxin (-0.4 V) is lower than that of NADP (-0.32 V), and the reaction proceeds directly. H^+ ions are pumped by the non-cyclic electron-transport system, but are accounted differently. Per pair of electrons transported, two H^+ ions are released from water in the lumen of the thylakoid, and two are taken up from the stroma, when the reduction of carbon dioxide takes place. Two more are taken across the thylakoid membrane by the reduction of plastoquinone (consuming H^+ ions from the stroma) and by its reoxidation at the *bf* complex (releasing them into the thylakoid lumen). In chloroplasts the Q-cycle probably does not operate (except possibly at low light intensities or when light is first switched on).

Cyclic electron transport makes use of PSI only. The ferredoxin that is reduced by PSI reduces plastoquinone (by means of an enzyme that is not known, but provisionally named ferredoxin-plastoquinone reductase, FQR). The electron transport from plastoquinone back to PSI is the same as in the non-cyclic scheme (Figure 5.5). The benefit is presumably the energy that is stored in the pumping of two H^+ ions per pair of electrons, associated with the reduction and oxidation of plastoquinone as described above. Observational support for the operation of a PSI-dependent system comes from the ability of cells, particularly algae, to carry out processes known to require ATP energy, such as the accumulation of glucose or K^+ ions from the medium. The cells require either oxygen for respiration, or light, and light that is of too long a wavelength to be efficiently absorbed by PSII is effective. The process is also seen to be immune to inhibitors of PSII such as the herbicide diuron, which blocks non-cyclic electron transport.

Cyclic electron transport in PSI is inhibited by antimycin A, although the site of inhibition is not known and is placed in the hypothetical FQR.

5.2.3 *Green sulphur bacteria*

Green sulphur bacteria are believed to carry out a cyclic electron-transport process virtually identical to the PSI-dependent version of green plants,

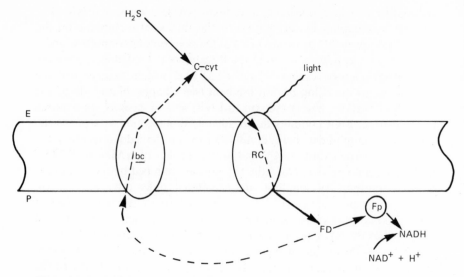

Figure 5.6. Overall photosynthetic electron transport in green bacteria. Electrons pass from hydrogen sulphide via a c-cytochrome to the reaction-centre (RC) and thence to ferredoxin and NAD. Note that NAD, not NADP, is the coenzyme. A non-cyclic pathway is shown; a cyclic path similar to that shown in Figure 5.5 may be considered likely. The bc complex is only required for the cyclic pathway. The c-cytochrome is not necessarily the same in the cyclic and non-cyclic pathways.

except that the quinone is menaquinone rather than plastoquinone. On the other hand, their non-cyclic electron transport is more similar to that of the purple bacteria, in that environmental reductants such as sulphide are used to reduce the cytochrome c-553 that is the electron donor to the photosystem (Figure 5.6).

5.3 Summary

Figure 5.7a represents electron transport in the chloroplast (compare Figure 4.14) in the well-known Z-scheme, showing the standard redox potentials of the intermediates. It illustrates the conservation of some 1.1 eV of redox energy in non-cyclic electron transport (in the products oxygen and NADPH) from an input of $2 \times 1.8\,eV$ of captured light quanta. Allowing for 2/3 ATP per electron transported, that is, 0.2 eV, the efficiency of conservation is 1.3/3.6 or 36%.

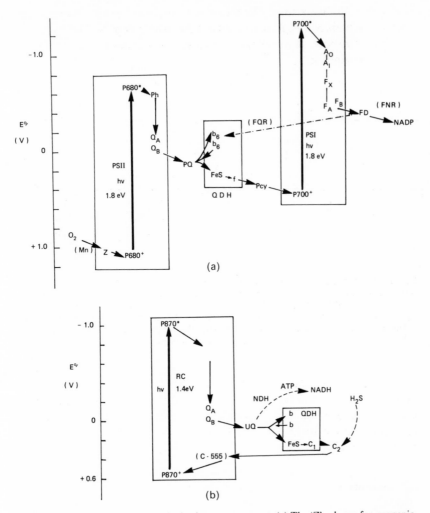

Figure 5.7. Diagrams of photosynthetic electron transport. (a) The 'Z' scheme for oxygenic systems e.g. chloroplasts. Redox intermediates are placed on an ordinate according to their standard potentials (see Table 6.1). FNR, ferredoxin:NADP (oxido)reductase; FQR, the hypothetical ferredoxin:plastoquinone reductase; QDH, the quinol dehydrogenase complex. The broken line represents the cyclic electron transport connection around PSI. H^+ ion translocation, that drives ATP formation, is associated with oxygen production, the reduction and oxidation of PQ, and the consumption of NADPH by metabolism. (b) Generalised purple bacteria. The scheme follows the principle of (a). The dashed lines show the formation of NADH at the expense of ATP (reverse electron transport) and of an environmental reductant such as hydrogen sulphide.

Figure 5.7*b* shows the cyclic electron transport of purple bacteria and their resemblance to PSII. Also shown is the reduction of NADH by reverse electron transport, at the expense of an environmental reductant such as hydrogen sulphide.

CHAPTER SIX

THE PRODUCTION OF ATP
Times from 1 s to 100 s

The history of the discovery of ATP and its involvement in the molecular mechanisms of biology is sketched in Table 6.1. The discovery, by Arnon *et al.* (1954), that isolated chloroplasts were able to produce ATP (photophosphorylation), was a major breakthrough in our understanding of plant cell biology, and the appearance of the chemiosmotic theory (Mitchell, 1961) had a similar impact in connecting areas of study which had been regarded as separate. In both cases furious debates were set in motion, generating an exhilarating torrent of experiment, before the new ideas became generally accepted.

6.1 Electron transport generates an H^+-ion gradient

In the electron-transport systems described above, it was noted that H^+ ions are moved or pumped from the cytoplasmic or stromal side of the membrane to the periplasmic side, or the lumen of the thylakoid. This is achieved by three mechanisms. The first depends on the role of quinones (ubi-, plasto- or menaquinone) as mobile redox carriers between protein complexes. Reduction of quinone to quinol, which takes place at the reaction centre (of purple bacteria, or green plant PSII) requires H^+ ions, taken from the P side of the membrane; when the quinol is re-oxidised by complexes of the *bc* type the H^+ ions are released on the opposite (E) side. Quinones also take part in the Q-cycle (more or less agreed upon in mitochondria and purple bacteria, but disputed in chloroplasts) around the *bc* complex, and here again H^+ ions are taken from the cytoplasmic side when quinone is reduced, and expelled on the periplasmic side when the quinol is oxidised.

The second mechanism depends on the metabolic consumption of H^+ ions in the cytoplasm and the production in the thylakoid lumen. The

Table 6.1 Development of our understanding of ATP production.

Date	Workers	Observation
10904	A. Harden and W.J. Young	Dialysis inhibits fermentation in yeast homogenates, hence soluble factors are required
1929	K. Lohmann C. Fiske and Y. SubbaRow	ATP isolated from muscle
1930s	O. Warburg O. Meyerhof H. Kalckar V. Belitser V.A. Engelhardt M.N. Lyubimova	ATP produced from glycolysis of glucose to lactic acid in muscle ATP produced from aerobic oxidations in animal tissues Myosin catalyses ATP hydrolysis
1941	F. Lipmann	Proposal that ATP is the central carrier of cellular energy
1943	S. Ruben	Proposal that ATP connected dark and light reactions of photosynthesis
1948	A. Todd et al.	ATP structure proved by synthesis
1953	E.C. Slater	Formulation of 'Chemical Intermediate' hypothesis for oxidative phosphorylation
1954	D.I. Arnon et al.	Discovery of photophosphorylation in isolated chloroplasts
1954	A.W. Frenkel	Photophosphorylation found in bacterial chromatophores
1957	Arnon et al.	Photophosphorylation related to light driven transfer of electrons from water to $NADP^+$: defined as non-cyclic, the 1954 observation a cyclic Coupling and control similar to mitochondrial oxidative phosphorylation
1961	P. Mitchell	Proposal that proton gradients (PMF) could connect electron transport with phosphorylation (Chemiosmotic Hypothesis)
1963	G. Hind and A.T. Jagendorf	Discovery of acid-bath phosphorylation in chloroplasts, supporting PMF principle
1967	E. Racker	Identification of coupling-factor (CF) with stalked 9 nm particles

consumption takes place when ferredoxin, reduced by PSI in green plants and green sulphur bacteria, provides electrons for the coenzyme-mediated reduction of CO_2. The production of H^+ ions results when water is oxidised to O_2 by PSII. Artificial reductants such as diaminodurene (DAD) or 2,6-dichlorophenol-indophenol (DCIP) also provide H^+ ions in this way when they pass through the thylakoid membrane and donate electrons to PSI on the luminal side, and by this means the experimenter can study electron transport, coupled to ATP formation, in PSI without the need to physically remove PSII. Obviously the protons consumed in the P-phase are not the same ones produced in the E-phase, and the mechanism can hardly be described as pumping, but the result is the same: some of the energy released by the redox reactions of electron transport is stored as an H^+ gradient.

The third mechanism for H^+ ion pumping is known as the conformational protein pump, which may turn out to be more important in the mitochondrial complexes I and IV (Figure 6.1) than in photosynthetic systems. It is the only explanation on offer for the action of the protein bacteriorhodopsin, in which excitation by light of the chromophore (all-trans-retinal) leads to conformational changes (detected by spectral changes in the retinal-protein) and expulsion of a proton from the cell (Figure 6.2). There is no detailed mechanistic molecular explanation for the action of bacteriorhodopsin. In the electron-transport complexes I and IV of the mitochondrion (and of purple bacteria growing aerobically) the conformational pump has been proposed, without however stating which of the possible polypeptides is performing the pumping, still less whether and how the mechanism is related to that of bacteriorhodopsin. Nevertheless, the concept of the conformational H^+-pump cannot be ignored.

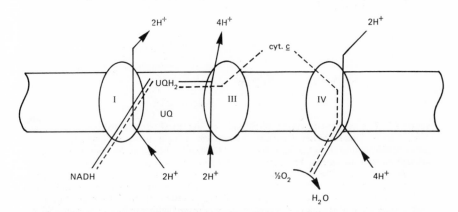

Figure 6.1. Proton pumping in mitochondrial respiration. The oxidation of NADH by $\frac{1}{2}O_2$ is shown here; the path of the electrons is indicated by dotted lines. The transfer of $8\,H^+$ ions from the P-side to the E-side is shown by solid arrows. One pair enter at complex I and are carried by UQH_2 to complex III (the *bc* complex), one pair is carried by the Q-cycle in complex III (see Figure 5.2), and one pair is consumed by the formation of water. The remaining pairs carried by complexes I and IV may be driven by a conformational-pump mechanism. See Nicholls (1982).

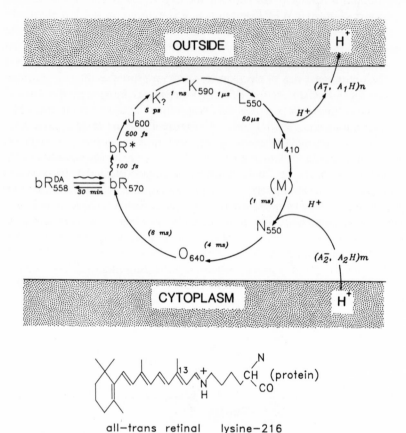

Figure 6.2. The energy-conserving photocycle of bacteriorhodopsin (bR) in the purple membrane of *Halobacterium halobium*. Absorption of light (wavy arrow) leads to the successive formation of bR*, J, K, L, M, N and O, shown with subscripts that indicate their absorption maxima in nm. The time-constants for reactions are shown ($1\,\text{ps} = 10^{-12}\,\text{sec}$ and $1\,\text{fs} = 10^{-15}\,\text{sec}$). The entities ($A^{-}$, AH) are dissociating groups in one or more proteins that convey the protons from the cytoplasm to the bacteriorhodopsin (at N), and from bacteriorhodopsin (L) to the outside of the cell, thus building up pH differences of up to 4 units. These changes are due to the isomerisation of retinal (attached by a protonated Schiff base to the ε-amino group of lysine-216—lower diagram) from the strained, all-*trans* isomer to the 13-*cis*-isomer and subsequently to other protein components. Also shown is the slow interconversion between bR and an inactive, dark-adapted from ('DA'). (Copied from an unpublished diagram provided by Professor W. Stoeckenius based on Mathies *et al.*, 1988, and Kouyama *et al.*, 1988.)

6.2 A proton gradient has both electric and pH components

It is not realistic to suppose that any region of (biological) space can contain a chemically significant imbalance in the total number of positive and negative charges. This is the principle of electroneutrality. If H^+ ions are pumped out of the cytoplasmic compartment, then a counter-movement must take place such that anions travel in the same direction, or another cation moves in the opposite direction. In these circumstances H^+-pumping leads to a pH difference across the membrane. If no counter-ionic movement is possible, then the pump will create a membrane potential, positive on the side towards which the H^+ ions are being pumped (i.e. the periplasmic side). In this latter case the membrane potential will, virtually immediately, prevent any detectable proton movement and hence there will be no observable pH change. If this is hard to visualise, imagine people in man–woman couples, determined to remain holding hands, when only one

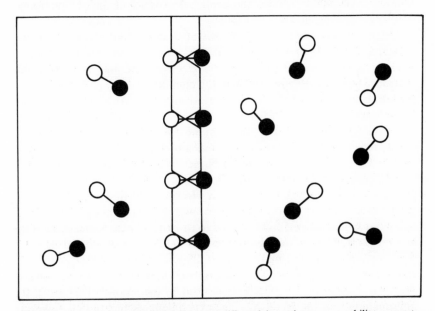

Figure 6.3. Ion-concentration differences, and differential membrane permeability, generate membrane potentials with virtually no net transport. The diagram shows a plan view of a barrier with turnstiles through which only one member of a couple may pass, showing that the stress (potential) is due to polarisation of the membrane by the orientation of a few couples, and not to any net inequality of (+) and (−) concentrations in any region. For the purpose of the illustration, the ions (literally 'wanderers') are shown attached in pairs; in solutions no such pairing would be apparent.

member of the couple is able to pass through a turnstile. The greater the pressure (from the crowd, or police, or whatever) the greater the tension developed by the couple holding hands across the turnstile. The turnstile (imagine a row of turnstiles in a wall) becomes polarised, men on one side, women on the other, with considerable energy stored in their pulling arms, but the areas on each side of the wall remain in man–woman balance (Figure 6.3).

In chloroplasts and purple bacteria, counter-ions such as Cl^- are able to penetrate the membranes of thylakoids and chromatophores so that a large pH difference builds up, and the high membrane potential, formed in the first millisecond of illumination, rapidly decays to much lower levels. The pH difference is due to a fall in the pH of the periplasmic/luminal space from 7 to around 5; in chloroplasts the stromal pH rises from 7 to about 8.5.

This is in contrast to the mitochondrion, where the inner membrane is virtually impermeable to chloride ions, and the pH difference is smaller. The membrane potential remains the principal product of the H^+-pumping resulting from the electron transport.

The combined effect of membrane potential and pH difference is to create a potential capable of doing work if H^+ ions are allowed to re-cross the membrane. Put a different way, the electric field and concentration difference will force H^+ ions through the membrane if any path exists. This potential, by analogy with the electromotive force (EMF), has been named the protonmotive force (PMF) by P. Mitchell, who developed this theory, known as the Chemiosmotic Theory, in the 1960s when it was believed that energy released in electron transport was stored in the form of a 'chemical intermediate'. Mitchell received the Nobel Prize for Chemistry in 1978.

Mitchell was at the time principally interested in the process of oxidative phosphorylation in the mitochondrion. However, the first item of evidence for the role of proton gradients came from experiments on chloroplasts by Hind and Jagendorf (1962). They tried to isolate the hypothetical 'chemical intermediate', by illuminating chloroplasts so that they performed electron transport, then transferring them to a different medium containing ADP, where they formed ATP. The chloroplasts appeared to be carrying a chemical substance from the first medium to the second. This energetic intermediate was known as 'X_E'. The experiment seemed to fail, however, when the control chloroplasts, which were not illuminated, produced as much ATP as the illuminated ones. In fact the ATP was produced whenever the pH of the first medium was lower by some 2 pH units than the second. 'X_E' was not a chemical substance but a pH difference across the thylakoid membrane. This experiment is now known as 'acid-bath phosphorylation'.

6.3 How much work can be stored by the proton gradient?

An H^+ ion possesses unit positive electric charge. If one mole of H^+ passes across a membrane under the influence of a membrane potential of E volts, then the energy released is equal to EF, where F is Faraday's equivalent or 96 400 Coulombs, and hence:

$$\text{electric work (max.)} = E \times 96.46 \text{ kJ.}$$

Fluids are subject to diffusion. Regions of unequal composition tend to become equal by the random movement of molecules. This is an application of the Second Law of Thermodynamics, that entropy (disorder) increases in a spontaneous process in a closed system. If we have a situation where selectively permeable membranes enclose regions of high solute concentration, solvent passes inwards so as to dilute the solute, generating the phenomenon of osmotic pressure; the consequent expansion of the membrane enclosure shows that work can be obtained from the process. The maximum amount of work that can be done when one mole H^+ (or anything) passes from a region of concentration c_1 to one of concentration c_2 is $RT \ln(c_2/c_1)$, where R is the gas constant ($8.314 \text{ J mol}^{-1} \text{ K}^{-1}$) and T the absolute temperature. This expression can be simplified: natural logarithms (ln) can be replaced by $2.303 \log_{10}$, and the negative logarithm of the hydrogen ion concentration is of course pH. Therefore, for one mole of H^+ passing from pH_1 to pH_2 the maximum work is:

$$\text{diffusion work} = (pH_2 - pH_1) \times 5.7 \text{ kJ (at 25 °C)}$$

$$\text{total (max. work)} = W_{max} = \text{electric work} + \text{diffusion work.}$$

Sign convention. The signs must be taken into account. It is normal practice to quote the membrane potential of a vesicle in terms of the inside potential with respect to the outside. Because of the sign convention in calculating pH, the pH difference is given as the external pH minus the internal pH. A positive membrane potential and a positive ΔpH both tend to drive H^+ ions out of the vesicle doing work given by:

$$W_{max} = 96.46 \times E_m + 0.59 \times \Delta pH \text{ kJ per mole } H^+.$$

W_{max} has been given the symbol Δp, and is expressed in $J \text{ mol}^{-1}$. Potentials, including protonmotive force (PMF), are expressed in volts:

$$PMF = W_{max}/F = E_m + 0.059 \times \Delta pH.$$

Work is accounted as negative when it is performed by a system on its surroundings (that is, the system loses energy). Biological membranes that conserve energy in ATP from electron transport or bacteriorhodopsin are topologically related; the membrane potential is always negative, and the pH higher, on the cytosolic, matrix or stroma side, with respect to the periplasmic, inter-envelope or thylakoid-lumen side, when ATP is being produced. The former is referred to by electron microscopists as a P- (protoplasmic) side, the latter as an E- (external) side. Whether these sides correspond to the inside or outside of vesicles depends on the particular system and its experimental preparation. It is therefore better from the point of view of bioenergetics (accounting for ATP formation) to adopt a sign convention such that PMF is negative when it will drive H^+ ions from the E-side to the P-side of a membrane.

6.4 The proton gradient drives ATP formation

6.4.1 *How much energy is required for ATP synthesis?*

ATP readily undergoes hydrolysis to ADP and P_i. The equilibrium between ATP and its hydrolysis products is overwhelmingly toward the latter, and the equilibrium constant for the reaction is approximately 10^5:

$$ATP \rightarrow ADP + P_i$$

$$K_c = \frac{[ADP][P_i]}{[ATP]} = 10^5 \, mol \, dm^3$$

where the square brackets denote concentrations. The reasons for the virtually complete hydrolysis apply to pyrophosphate bonds in general, and indeed to any compound of the anhydride type. First, the products of hydrolysis are weak oxyacids that are stabilised by resonance. Secondly, the phosphate groups of pyrophosphates are negatively charged, and repel each other electrostatically. Thirdly, there is a pH-dependent heat of dissociation of the acid groups formed by the hydrolysis, and fourthly, there is a gain in the entropy of the system owing to the increase in the total number of molecules.

A system that is displaced from equilibrium can do work. The maximum quantity of work that is obtained when the exchange proceeds reversibly is given by the Gibbs energy change, ΔG. ΔG is calculated from, first, a knowledge of the standard Gibbs energy change, ΔG^0, that describes the reaction, and the actual concentrations of reactants and products that describes the system itself. ΔG^0 is calculated from the equilibrium constant K_c:

$$\Delta G^0 = -RT \ln K_c:$$

for ATP, $\quad \Delta G^0 = -RT \ln(10^5) = -28\,\text{kJ at }25\,°C.$

The negative sign indicates that work is done and heat is given out. The standard Gibbs energy is in fact the maximum work for a change when all relevant materials are in their standard states. The standard states are unit concentration ($1\,\text{mol dm}^{-3}$) for ATP, ADP and P_i, and using the 'biologists' convention' water is always standard and the pH is 7 or a stated value. The actual Gibbs energy change (ΔG) is a function of the concentrations of ATP, ADP and P_i and is given by:

$$\Delta G = \Delta G^0 + RT \ln[\text{ATP}]/[\text{ADP}][P_i]$$

$$= \Delta G^0 + 5.7 \log[\text{ATP}]/[\text{ADP}][P_i] \text{ at }25\,°C, \text{ in kJ mol}^{-1}.$$

Measurement of the concentrations of free (unbound) nucleotides in cells is not easy, but estimates are such that ΔG for ATP formation is expected to be in the range 50–$60\,\text{kJ mol}^{-1}$. This much energy must be provided by the PMF. There is a problem in that it is difficult to measure the number of H^+ ions that are required to pass through the F-ATPase system, for one ATP to be synthesised. A consensus view is that the number is three. The PMF must therefore amount to at least $20\,\text{kJ mol}^{-1}\,H^+$. This may be made up in various ways, as shown in Table 6.2.

Direct evidence for the existence of the PMF requires observation of the

Table 6.2 Composition of the PMF in various organelles.

Organelle	ΔpH	E_m	PMF (in volts)	Electrochemical potential (in kJ mol^{-1})
			$0.059 \times \Delta\text{pH} + E_M$	$5.7 \times \Delta\text{pH} + 96.48 \times E_M$
Chloroplast	-3.5	-0.05	$-0.206 - 0.05 = -0.256$	$-19.95 - 4.82 = -24.77$
Mitochondrion	-1.4	-0.14	$-0.086 - 0.14 = -0.226$	$-7.98 - 13.51 = -21.41$
Chromatophores	-1.0	-0.2	$-0.059 - 0.2 = -0.259$	$-5.7 - 19.3 = -25.0$

Values for the chloroplast and chromatophore are 'steady state': when the light is first turned on, values of E_M of 0.4 V or more are observed, which decay as the ΔpH builds up. Values for mitochondria are quoted from Stryer (1988) p. 411, and for chromatophores the discussion by Wraight et al. (1978) is most valuable. The sign convention is that ΔpH is $\text{pH}_P - \text{pH}_E$, and E_M is $E_E - E_P$, where the subscripts refer to E and P phases. Note that in mitochondria the inside is the P-phase, whereas the inside of thylakoids and chromatophores is an E-space. Nevertheless, when protons pass across the membrane, work is done in all cases which can be harnessed to the synthesis of ATP, and the PMF (etc.) is necessarily negative.

pH difference and membrane potential across the membranes of chromato-phores and thylakoids. It is straightforward to demonstrate that chromato-phores, or thylakoids performing cyclic phosphorylation, develop a change of absorption at about 515 nm. This is due to a physical effect on a carotenoid molecule in the reaction centre (PSII) caused by the electric field of the membrane potential (an electrochromic shift or Stark effect). The magnitude of the shift is proportional to the field, and can be calibrated. The field rises in two parts, a first part of some 50–100 mV, rising in picoseconds, due to the primary charge separation in the reaction centre, and a second part rising in milliseconds due to electron transport between quinone and the quinol reductase complex. In chromatophores the potential can reach 400 mV (positive in the E-phase, inside), but it declines in a few seconds as chloride ions penetrate, allowing the accumulation of HCl inside the vesicle.

The measurement of the internal pH is more difficult. It is easy to detect the uptake of H^+ from the medium, but more difficult to show that the ions enter the lumen, and how much buffering there is inside. Acceptable values were obtained by means of fluorescent indicators, or by measuring the kinetics of the reduction of cytochrome f, which are pH-dependent. They indicate that the luminal pH falls to about 5, while the stroma of the chloroplast rises to pH 8.5.

Both in bacteria and thylakoids there appear to be leaks; one is dependent on the pH difference, and the other on the P-phase pH. The latter is associated with the ATP synthase and is blocked by ATP or by ATP synthase inhibitors such as Dio-9. The effect of leaks is to prevent the field or pH difference reaching a magnitude that could damage the membrane. When ATP is being actively formed the leaks are not likely to be significant, so the rate of electron transport in the absence of ADP should not be subtracted in the presence of ADP, in arriving at values of the P/O ratio.

6.4.2   The ATP synthase or F-ATPase

ATP formation takes place in a membrane-bound particle known as the F-ATPase. This particle (distinct from the reaction-centre complexes and the *bc* complexes) is seen in negatively stained electron-micrographs as a knob, 9–14 nm in diameter, that projects from the membrane on the P-side. The knob itself, if detached, has Mg^{2+}-dependent ATPase activity, and is known as F_1 in the cases of mitochondria and bacteria, and CF_1 in the case of chloroplasts. The ATPase activity is latent (apparently inactive) in the chloroplast enzyme, and needs to be activated by thiol reagents, or by experimental treatments that partially disorganise protein complexes. The same is true of the ATPase activity of the intact thylakoid membrane, except that the natural activation process requires energy in the form of ATP or a high PMF across the membrane. The significance of the activation is that the thylakoid is inactive at night, and the ATPase needs to be switched off in order to prevent the reverse reaction taking place, consuming ATP and creating an unwanted PMF that dissipates. Mitochondria never rest! Bacteria, again, do not switch off their ATPase because they rely on a constantly maintained PMF for the uptake of amino acids, sugars, etc., from the medium.

The point of attachment to the membrane is a protein complex known as F_0 (CF_0). The complete system (F_0F_1) is the ATP synthase. If F_1 is detached, the membrane becomes freely permeable to H^+ ions. This permeability is mediated by F_0; if F_0 is purified and added to artificial (liposome) membranes, they become permeable to H^+.

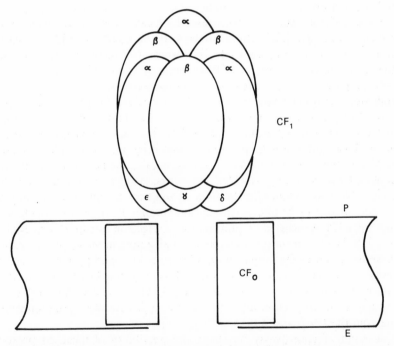

Figure 6.4. The subunit structure of the ATP synthase particle. The particle is known as F_1 in mitochondria and bacteria, and CF_1 in chloroplasts. It is attached to F_0, CF_0, respectively. The subunit structure of CF_0 is not shown. The diameter seen in electron micrographs with negative staining is 9 nm, a considerable underestimate of its true 14 nm.

F_1 is composed of five types of polypeptide, $\alpha, \beta, \gamma, \delta$ and ε. There are probably three copies of α and β in the particle which has a mass of some 400 kDa (Figure 6.4). The catalytic site is believed to reside on the β subunit. The α subunit binds nucleotides (such as ADP) and is known to undergo conformational changes. The γ subunit possesses SH groups (hence the need for thiol reagents in the experimental activation process).

F_0 contains three types of subunit, I, II and III; there are many copies of each. Subunit III contains the site sensitive to the energy transfer inhibitors. Both CF_1 and CF_0 contain some polypeptides coded in nuclear DNA, others coded in chloroplast DNA.

When protons are forced by a PMF to pass through the ATP synthase particle from the E-side to the P-side of the membrane, ATP is formed from ADP and P_i. If ATP is present, and there is no PMF, then protons will be driven from the P-phase to the E-phase at the expense of ATP, that is, a PMF will be created. (In the case of the chloroplast ATP synthase, it must be activated first.)

These activities can be demonstrated by means of liposomes, which are vesicles made of artificial lipid bilayers. If the complete ATP synthase particle is purified and added to liposomes, it is incorporated into the membrane of the liposome. If ATP is supplied, it is possible to demonstrate H^+ pumping.

Bacterial chromatophores take up H^+ in the dark if supplied with ATP.

If a PMF is created in the ATP synthase-loaded liposomes, for example by additionally incorporating bacteriorhodopsin into the membrane, then the ATPase synthesises ATP provided that it is supplied with light, ADP and P_i. All the actions of the ATP synthase require magnesium ions, and are inhibited by the 'energy-transfer' inhibitors, such as oligomycin, Dio-9, DCCD (N, N'-dicyclohexylcarbodiimide) and triphenyltin, as is the natural production of ATP by photosynthetic membranes. It is therefore concluded that the PMF, generated by photosynthetic electron transport, drives H^+ through the membrane (presumably through a pore in the F_0 particle) and in so doing causes the formation of ATP from ADP and P_i. The process is reversible; if the PMF is insufficient, ATP is hydrolysed by the F_1 particle and pumps protons so as to increase the PMF. Provided the ATPase is located in the membrane of a closed vesicle, it catalyses an equilibrium between the PMF and ATP.

A point of controversy has been the possibility of localised proton channels, so that H^+ ions may pass from a redox complex directly into an ATP synthase particle without passing through and therefore equilibrating with the bulk phases. This is argued from observations that ATP synthesis may (sometimes) commence before the bulk-phase pH difference has reached a sufficient value.

6.5 The PMF controls the rate of electron transport

In chloroplasts, mitochondria, chromatophores and other bacterial membranes, the rate of electron transport depends on the ability of the system to synthesise ATP, provided of course that the availability of light and the external oxidants and reductants are not limiting. The curve shown in Figure 6.5 shows an experiment measuring the production of oxygen (the Hill reaction) indicating electron transport. The initial rate of oxygen production is relatively slow, and is considerably increased by the addition of ADP ('State III', terminology of Chance and Williams, 1955). The ADP is used up (it becomes ATP), and the rate of change of oxygen concentration falls back to its previous value ('State IV'). The ADP effect can be repeated. This is termed 'photosynthetic control', by analogy with respiratory control by ADP in mitochondria. The rate is increased to its maximum by the addition of CCCP (carbonyl cyanide m-chlorophenylhydrazone). The term 'coupling' is used to indicate that electron transport is coupled to phosphorylation (phosphorylation here means ATP production) as a locomotive is coupled to a train. CCCP is an uncoupling agent (see below).

Spectroscopic studies of the components of electron-transport chains have indicated that certain steps are affected by the build-up of the PMF. In the chloroplast it appears to be the reduction of cytochrome f by plastoquinone (mediated by the Rieske FeS centre which cannot however be directly observed in this way). The rate of this reaction is effectively diminished as the pH falls from 7 to 5. The reactions of plastocyanin with P700 and with cytochrome f have also been suggested to be PMF control points.

The phenomenon of uncoupling can be brought about by any treatment that allows the PMF to dissipate. In the case of chloroplasts and chromatophores, where the PMF is chiefly due to a ΔpH, uncoupling reveals H^+ ions crossing the membrane by means other than the ATPase. Such means are (a) detaching CF_1 from CF_0 so that the membrane becomes permeable to protons, (b) providing lipid-soluble weak acids such as CCCP, which combine with protons rendering them soluble in the lipid phase, and (c) (not strictly the same process) the effect of an amine such as ethylamine or ammonia, which crosses the membrane by means of the lipid solubility of the free base, and then combines with protons forming the ammonium derivative which is positively charged and impermeant.

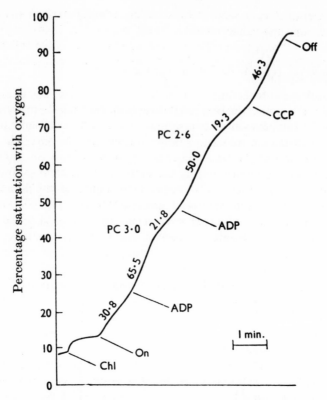

Figure 6.5. The PMF controls the rate of electron transport. Experiment by West and Wiskich (1968) showing polarographic (oxygen electrode) tracing when chloroplasts (264 μg Chl) were illuminated in a medium (4.4 ml) containing P_i at pH 7.2 with 2.3 mM $K_3Fe(CN)_6$ as electron acceptor. Addition of ADP (422 nmol) caused acceleration of oxygen production, followed by deceleration when the ADP was exhausted (state III/state IV transition). The photosynthetic control ratio (PC) is the ratio of the rates in state III and state IV. CCP (carbonyl cyanide m-chlorophenylhydrazone), an uncoupler, collapsed the PMF and allowed the maximum rate of electron transport (but note some deterioration during the experiment). The numbers along the trace are μM O_2 min^{-1}. Air-saturation was 250 μM oxygen (1.1 μmol).

In case (a) protons will move until the PMF is zero, that is, the sum of the components due to ΔpH and E_M is zero. Note that ΔpH and E_M may still have significant magnitudes. In the last case (c) the energy is still there but unavailable to the ATPase.

In some bacterial chromatophores a significant part of the PMF is due to the membrane potential, and then uncoupling can be achieved by means of *ionophorous antibiotics* such as valinomycin, which makes a pore in the membrane that specifically allows K^+ ions to pass. If K^+ ions are present, they will pass through the pore until the membrane potential collapses to zero. Note that the effects on ΔpH and the membrane potential must be considered separately.

The P/O ratio. Non-cyclic electron transport in chloroplasts is expected to result in four H^+ ions being moved across the thylakoid membrane for every oxygen atom released as gaseous oxygen (or for every $2e^-$ transported): two produced from water oxidation in the lumen, two consumed from reduction of carbon dioxide in the stroma, and two transported by the plastoquinol/quinone loop. This neglects any contribution from the Q-cycle. Given a likely need for three H^+ per ATP formed by the synthase, the ratio of ATP formed to oxygen released is expected to be 4/3, in good agreement with experimental findings.

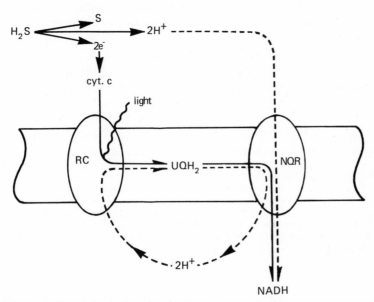

Figure 6.6. Reverse electron transport. The scheme of Figure 5.4(a) and (b) is adapted to show that the reduction of NAD at the expense of hydrogen sulphide in purple sulphur bacteria (on the right) may be driven by a proton gradient (PMF) set up by cyclic electron transport (on the left). The reduction of NAD is inhibited by uncoupling agents, which dissipate the PMF, unlike the situation in the green bacteria (Figure 5.6). There is some indication that a reverse electron transport process, reducing NAD by means of PQH_2, may be possible in chloroplasts (Godde, 1982).

The effect of uncouplers is to reduce the P/O ratio while stimulating electron transport (given saturating light). The effect of ATP-synthase inhibitors is to slow both electron transport and ATP formation, preserving the P/O ratio.

6.6 Reverse electron transport

Photosynthetic electron transport is not a reversible process in any organism because so much of the energy from the light is lost as heat. Nevertheless, some individual steps in mitochondrial electron transport and in purple bacteria are reversible, in particular between the coenzyme NAD, the complex NADH-ubiquinone reductase (complex I of mitochondria), ubiquinone and the bc complex (complex III of mitochondria). If the PMF is sufficiently high, electrons can be driven from the bc complex into NADH (Figure 6.6). This has no biological significance in mitochondria, and only serves as an experimental demonstration of the importance of the PMF in the chemiosmotic theory. In all the groups of non-oxygen-producing photosynthetic ('anoxygenic') bacteria however, it allows the reduction of NAD^+ to NADH at the expense of relatively weak reductants, by means of the energy present in the PMF generated from photosynthetic electron transport (see p. 88). Weak reductants include succinate, acting via a succinate-ubiquinone reductase (like complex II of mitochondria) or sulphide (see Figure 5.7), acting via a H_2S-cytochrome reductase and cytochrome c in the periplasmic space and reducing ubiquinone via the photosynthetic reaction centre.

CHAPTER SEVEN

METABOLIC PROCESSES AND PHYSIOLOGICAL ADJUSTMENTS
Seconds to hours

The turnover time of enzymes, when saturated with their substrates, may be of the order of 10^{-5} to 0.1 s. This time is taken up by the need of the protein to undergo changes in conformation. In the cell, however, enzymes are virtually never saturated with substrate, indeed the concentration of an enzyme may exceed that of its substrate. The rate of metabolic processes is strongly dependent on diffusion, so that the actual turnover times of enzymes may be measured in seconds, and complete pathways may take 10–100 s to adjust to changes in the supply of feedstock.

In this chapter we shall describe some significant enzyme-catalysed processes of photosynthesis. The most massively important process is of course the nutritional pathway of the organism, such as carbon dioxide fixation in green plants, but the set of reductions that depend on ferredoxin are not to be overlooked.

7.1 Ferredoxin-dependent reactions

Ferredoxin, reduced by PSI of green plants, has two obvious roles, (a) of reducing NADP via a membrane-bound FAD-containing enzyme FNR, and (b) of reducing plastoquinone via a membrane-bound reductase FQR that is at present unknown. Both activities may be associated with the $b_6 f$-complex.

It is important not to overlook the other roles of ferredoxin. These include reduction of inorganic materials such as sulphate, nitrite, nitrogen, hydrogen and oxygen, of proteins such as thioredoxin and of organic compounds such as acetylene and (in bacteria) synthesis of 2-oxoacids (Figure 7.1).

Sulphate ions enter the chloroplast probably by means of a specific transporter and are activated by ATP in stages giving adenosine

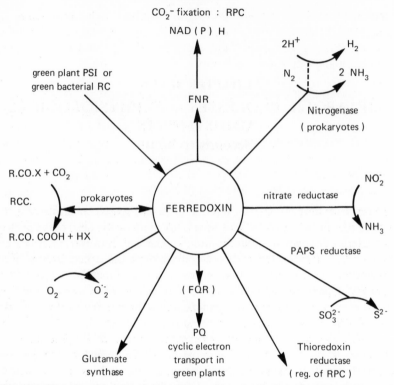

Figure 7.1. The roles of ferredoxin. The diagram shows ferredoxin reduced by PSI in green plants or by the green-bacterial reaction centre, or by the phosphoroclastic reaction in other bacteria. In turn ferredoxin reduces the other materials as shown. Note that production of superoxide or hydrogen is not of any value. FNR: ferredoxin: NADP oxidoreductase. FQR: ferredoxin:quinone reductase (hypothetical).

phosphosulphate (APS) and phosphoadenosine phosphosulphate (PAPS). The sulphate group is attached to a thiol group of an (unidentified) protein, at which point it is equivalent to sulphite. The reduction to sulphide requires six reducing equivalents from ferredoxin, and is catalysed by thiosulphonate reductase. Sulphide is principally used for the production of cysteine.

Nitrate is the main nitrogen source for higher plants, and is reduced to nitrite in both roots and leaves. The enzyme nitrate reductase uses electrons from NADPH, and has three prosthetic groups that form an electron-transport chain: FAD, cytochrome b-557 and a molybdenum atom. Nitrite (a toxic substance) enters the chloroplast and is reduced to ammonia by either sulphite reductase or nitrite reductase; again, six equivalents of ferredoxin are required. Both enzymes contain iron–sulphur centres (Fe_2S_2 in the case of nitrite reductase, Fe_4S_4 in the case of sulphite reductase) and one sirohaem group.

Nitrogenase, which fixes nitrogen from the air into ammonia, is the largest example of an iron–sulphur system, and is found in prokaryotes, the non-photosynthetic root-nodule bacteria (*Rhizobium*) of leguminosae, many species of green and purple bacteria, and many blue-green bacteria, of which the filamentous *Anabaena cylindrica* is a well studied example. Some of the cells are modified in shape and size and are known as heterocysts; they contain PSI but no PSII and possess thick cell walls; these features avoid the production of oxygen within the heterocyst and restrict its inward diffusion. Molecular oxygen destroys nitrogenase. The electron donor to PSI may be glucose-6-phosphate, via cytochrome *c*. There are two protein components, the first carries 2 Mo and 24–32 FeS, while the second carries an Fe_4S_4 centre. In addition to six equivalents of reduced ferredoxin, some 24 ATP are also required per nitrogen molecule. Two molecules of ammonia are formed. In the absence of nitrogen, water or H^+ ions are reduced to hydrogen, hydrogenase activity. (This is not the same hydrogenase system found in algae that do not possess nitrogenase.) Acetylene (C_2H_2) is an experimental substitute for nitrogen, used in the assay of the nitrogenase system: ethylene is produced, detected by gas chromatography.

Ammonia is incorporated into amino acids in the chloroplast stroma, first forming glutamine from glutamic acid and ATP, following which the amide group is transferred by glutamate synthase to 2-oxoglutarate, using reduced ferredoxin:

$$\text{gln} + 2\text{OG} \xrightarrow{\;2e^-\,(FD)\;} 2\,\text{glu}.$$

During the operation of photorespiration in the cells of C3 plants (see below), ammonia is liberated in large amounts, and an expenditure of ATP and reduced ferredoxin is required to re-incorporate it by the above process.

Thioredoxin is a small protein containing two thiol groups that can be reversibly oxidised. It is reduced by ferredoxin, by means of the enzyme thioredoxin reductase, which is itself another FeS protein. Thioredoxin is important in regulating certain enzymes such as fructose bisphosphatase by maintaining their sulphur in the reduced (SH) state. It also (probably) reduces the γ subunit of the F-ATPase during activation.

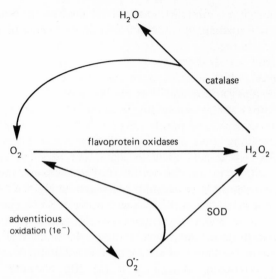

Figure 7.2. The roles of catalase and superoxide dismutase. Hydrogen peroxide is produced by flavoprotein oxidases (see Figure 7.5). O_2^- is produced by reduction of oxygen (see, e.g. Figure 7.1). Superoxide dismutase (SOD) forms oxygen and hydrogen peroxide, and catalase forms oxygen and water. The pair therefore remove the dangerous oxides.

Molecular oxygen is known to be reduced *in vitro* by a one-electron reaction with reduced ferredoxin, forming superoxide (O_2^-). In the cell there is a high activity of the enzyme superoxide dismutase, which disproportionates superoxide to oxygen and hydrogen peroxide, and of catalase which decomposes hydrogen peroxide to oxygen and water (see Figure 7.2). It is therefore likely that superoxide is formed *in vivo* and continuously scavenged; superoxide is regarded as a very destructive entity, even though its lifetime in the cell is less than 1 ms.

Ferredoxin, in the green sulphur bacteria, drives the reductive carboxylation of acyl coenzyme A derivatives:

$$CO_2 + 2e^- FD + \overset{\displaystyle CH_2COO^-}{\underset{\displaystyle \underset{\text{succi.yl CoA}}{CH_2CO-SCoA}}{|}} + H_2O \xrightarrow{\text{ferredoxin}}$$

$$\overset{\displaystyle CH_2COO^-}{\underset{\displaystyle \underset{\text{α-oxoglutarate}}{CH_2CO\cdot COO^-}}{|}} + \underset{\text{coenzyme A.}}{CoASH} + H^+$$

The significance of this reaction is that it enables the bacteria to live photoautotrophically, fixing carbon dioxide by means of the reductive citrate cycle (below).

7.2   Carbon dioxide fixation

7.2.1   *The reductive citrate cycle*

The reductive citric acid cycle (RCC) found in some green sulphur bacteria is not of any massive significance in the biosphere, but it is of considerable interest. It is the only alternative to the fixation of carbon dioxide by means of the reductive pentose cycle (RPC, below), found in green plants, algae and purple bacteria. The reactions of the RCC shown in Figure 7.3a, apart from the reductive carboxylation of acyl CoA (above), are the same as the oxidative citrate cycle (OCC) also known as Krebs' cycle. The OCC (Figure 7.3b) is the principal metabolic pathway of aerobic respiration in most aerobic bacteria, and mitochondria in most eukarytic cells including animals, algae and green plants. The OCC completely oxidises material at the level of acetate to carbon dioxide and water, driven by the oxidising conditions produced by molecular oxygen and the respiratory electron-transport chain. The severely reducing conditions brought about by the photosynthetic electron transport chain in the green bacteria together with the supply of ATP are sufficient to drive the RCC in the direction of production of acetate.

The acetyl CoA produced by the RCC is reduced and carboxylated by the ferredoxin-dependent reaction described above, generating pyruvate. Pyruvate is converted to phosphoenolpyruvate by an unusual, direct reaction in which both terminal phosphate groups are lost from ATP (compare the corresponding step in the C4 pathway, section 7.2.4).

7.2.2   *The reductive pentose cycle*

In spite of the variety of reactions that consume photosynthetic reducing power, the balance of carbon dioxide uptake and oxygen output is very close to unity in green tissues and algal cultures. Virtually all the reduced ferredoxin is used to reduce NADP via the enzyme FNR, and the NADPH is consumed in the reductive fixation of carbon dioxide. This takes place within the chloroplast by means of the reductive pentose cycle (RPC). The RPC is also the basis of carbon dioxide fixation by the purple bacteria (Chromatiaceae) and the chemolithotrophic bacteria. Some of the

Rhodospirillaceae can also use the cycle, for either photosynthetic or chemolithotrophic carbon dioxide fixation.

Overview. Carbon dioxide is reduced by means of the RPC, in the chloroplasts of green plants, to the level of carbohydrate; the chloroplast exports triose phosphates to the cytoplasm, and builds up a temporary store of starch from glucose units when the rate of carbon dioxide fixation

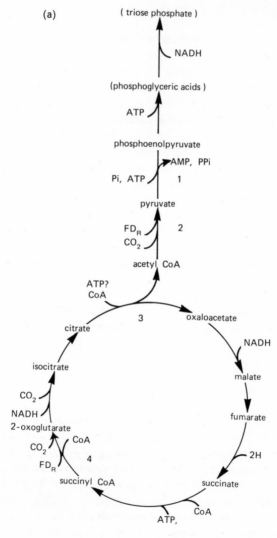

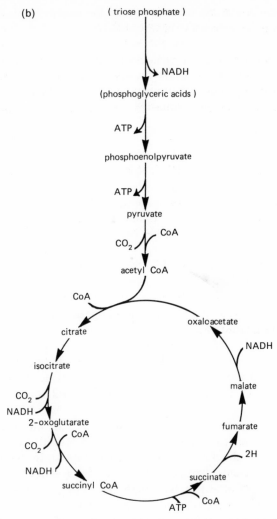

Figure 7.3. (*a*) The reductive citrate cycle. This pathway is practised by only a few species of green sulphur bacteria, but it is the only alternative to the reductive pentose cycle for photosynthetic carbon dioxide fixation. It is shown in a version for the synthesis of carbohydrate, but was originally perceived as a means for the synthesis of amino acids such as alanine, aspartate and glutamate. Note the four reactions which differ from the equivalent in the oxidative citrate cycle: 1, pyruvate, P_i, dikinase; 2, pyruvate synthase; 3, the citrate cleavage enzyme or citrate lyase and 4,2-oxoglutarate synthase. FD_R: reduced ferredoxin. (*b*) (For comparison) the oxidative citrate cycle. This is the principal energy-yielding oxidative pathway of mitochondria and aerobic bacteria. Most of the enzyme-catalysed reactions are the same as in the reductive cycle.

exceeds the rate of utilisation of triose in the cytoplasm. Most of the triose in the cytoplasm is converted to hexose and thence to sucrose, that is temporarily stored in the vacuole (of the leaf cell) and, later, translocated via the conductive tissue to other parts of the plant. Some of the reactions of the RPC take part in a shuttle system which effectively provides ATP for general use in the cytoplasm; it appears that the mitochondria in the same cells are switched off, so that respiration does not take place simultaneously with photosynthesis.

Brief history. The 'biochemical phase' of photosynthesis research began with the elucidation of the RPC in M. Calvin's laboratory at Berkeley, California in the years up to 1951. The work depended on the technique of radiochemical labelling, in this case using the isotope ^{14}C. The second essential was the resolution of organic substances by means of paper chromatography, developed in A.J.P. Martin's laboratory in 1945. In this way 3-phospho-D-glyceric acid (3PGA) was identified as the first detectable radioactive material formed when algal cells (of *Chlorella*) were allowed to perform photosynthesis in the presence of $^{14}CO_2$ for times down to a few seconds; the cells were killed, and the metabolites were extracted in methanol and resolved on a paper chromatogram. The radioactivity was detected by radioautography on X-ray film placed in contact with the paper. The identities of the spots were determined by means of known standards. The ^{14}C first appeared in C-1 of the PGA, but at longer times the other carbon atoms were labelled. This was evidence that the precursor of the PGA was formed from the PGA itself, and hence a cycle existed.

The RPC, also known as the Calvin–Benson cycle, was deduced from the experiments described above, even though several crucial enzymes were not shown at the time to exist in the chloroplast. In this as in many other examples science has progressed not by a series of steps made logically secure one at a time, but by an imaginative proposal supported but not proved by experimental evidence, that was recognised as an improvement on previous ideas on the subject, and for which the logical proof was only completed many years later when the topic had somewhat lost its novelty.

Details. The unique feature of the reductive pentose cycle is the enzyme known as RuBisCO (ribulosebisphophate carboxylase/oxygenase, or 3-phosphoglycerate CO_2-lyase, dimerising, E.C. 4.1.1.39.). It catalyses the reaction whereby ribulosebisphosphate (Ru15BP) is carboxylated by means of carbon dioxide:

$$Ru15BP + CO_2 \rightarrow 2 \ 3PGA.$$

The six-carbon intermediate is 2-carboxyarabinitol-1,5-bisphosphate. Unlike the several carboxylations of the reductive citrate cycle, this is the only point at which carbon dioxide is taken up in the RPC.

The enzymes of the cycle are shown in Figure 7.4. ATP is required for phosphorylation of 3PGA to 1,3-bisphosphoglyceric acid (BPGA) by means of the enzyme phosphoglycerate kinase. This reaction is reversible (that is, it is close to equilibrium) in both chloroplast and cytoplasm.

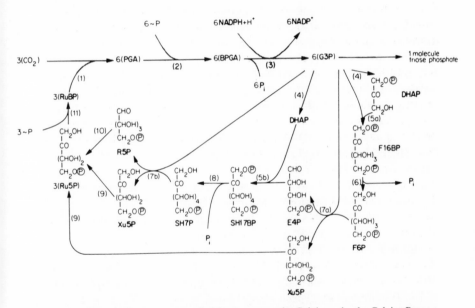

Figure 7.4. The reductive pentose cycle (also known as the Calvin cycle, the Calvin–Benson cycle and the photosynthetic carbon reduction (PCR) cycle). The shaded portion shows the fixation of three molecules of carbon dioxide into triose phosphate (the principal export from the chloroplast). Abbreviations for metabolites: BPGA, 1,3-bisphospho-D-glyceric acid; DHAP, dihydroxyacetone phosphate; E4P, D-erythrose-4-phosphate; F16BP, D-fructose-1,6-bisphosphate; F6P, fructose-6-phosphate; GAP, D-glyceraldehyde-3-phosphate; P_i, inorganic phosphate; P, combined phosphate; R5P, D-ribose-5-phosphate; RuBP, D-ribulose-1,5-bisphosphate; SH17BP, D-sedoheptulose-1,7-bisphosphate; SH7P, sedoheptulose-7-phosphate; Xu5P, D-xylulose-5-phosphate. Enzymes: 1, RuBisCO; 2, phosphoglycerate kinase; 3, GAP reductase (NADP-linked); 4, triose phosphate isomerase; 5a and 5b, aldolase; 6, fructose-1,6-bisphosphate 1-phosphatase; 7a and 7b, transketolase; 8, sedoheptulose-1,7-bisphosphate 1-phosphatase; 9, pentose-5-phosphate 3-epimerase; 10, ribose-5-phosphate isomerase and 11, phosphoribulokinase.

BPGA is reduced to glyceraldehyde-3-phosphate (GAP) by means of NADPH and the enzyme GAP reductase (NADPH-linked, as opposed to the similar but NAD-linked enzyme in the cytoplasm that is chiefly involved in glycolysis). GAP and the ketotriose, dihydroxyacetone phosphate (DHAP) are interconverted by triose phosphate isomerase; the equilibrium is in favour of DHAP. DHAP is the principal form of triose exported to the cytoplasm.

The remaining reactions may be regarded as a means of regenerating the original material, Ru15BP. However this would be to assume that triosephosphate is always the product exported to other pathways. In fact the RPC is autoanaplerotic—self-balancing—and can adjust itself so that any intermediate can, in theory, be output as product. To continue, DHAP and GAP take part in the aldol condensation catalysed by *aldolase* to form fructose-1,6-bisphosphate (F16BP), and the equilibrium is in favour of the hexose. The bisphosphate is hydrolysed by a *phosphatase* (at the C-1 position) yielding fructose 6-phosphate (F6P). The enzyme *transketolase* then transfers C-1 and C-2 (the ketol group) from F6P to GAP, forming the ketose xylulose 5-phosphate (Xu5P). C-3 to C-6 of the original F6P are left as the aldotetrose, erythrose 4-phosphate (E4P), to which a specific *aldolase* adds DHAP forming the sedoheptulose 1,7-bisphosphate (SH17BP). The same 1-phosphatase forms sedoheptulose 7-phosphate, as in the case of F16BP. Transketolase transfers the ketol group from SH7P to G3P, leaving C-3 to C-7 in the form of the aldopentose ribose 5-phosphate (R5P), and forming Xu5P as before. Xu5P has the same C-3 conformation as F6P and SH7P, and all the transketolase reactions are reversible. The ketol group is carried on the coenzyme thiamine pyrophosphate. Xu5P is converted to ribulose 5-phosphate (Ru5P) by means of *Xu5P:3-epimerase*, and R5P is converted to Ru5P by the enzyme *phosphoriboisomerase*. Finally, ATP is required for the phosphorylation of Ru5P to RuBP by *phosphoribulokinase*.

All the above enzymes have now been found in the chloroplast stroma in sufficient quantities to account for carbon dioxide fixation.

It will be apparent that the formation of one mole of DHAP requires three CO_2, nine ATP and six NADPH. The ratio in which ATP and NADPH are produced in the light-reactions of photosynthesis (the non-cyclic process) is probably 1.33:1; if this is correct then an additional source of ATP is needed to support the reductive pentose cycle, and it presumably comes from cyclic electron transport around PSI.

RuBisCO is a fascinating enzyme, with many interesting and important properties. It is the most abundant protein in nature. This is because

(i) virtually all the carbon dioxide incorporated into the biosphere passes through RuBisCO, and (ii) of all the enzymes of the RPC, it has a slower catalytic rate than most of the many enzymes which produce carbon dioxide, and more molecules of it are required. The molecules are (per catalytic site) relatively large in mass. RuBisCO is found to be at least half of the protein of the chloroplast, and the chloroplasts contain more than half the protein of a leaf. In many prokaryotic cells RuBisCO is found in 'polyhedral bodies', and it has been suggested that is is also a convenient food-storage protein. The pyrenoids of algal chloroplasts are largely made up of RuBisCO.

As a protein it is complex and complicated, having in higher plants eight large (A) subunits and eight small (B) subunits. The B subunits are synthesised in the cytoplasm, and the B-precursor enters the chloroplast by means of a transporter protein located in the chloroplast envelope. The precursor is cut down to size by proteinases, and the subunits are assembled by means of a binding protein. One bacterium, *R. rubrum*, has only A subunits (hence A is the catalytic centre), and in some other bacterial species A_2B_2 forms are known. A_8B_8 is universal in green plants.

The catalytic rate of an enzyme-catalysed reaction depends on, among other things, the concentration of substrate, reaching a maximum (saturation) when the substrate concentration is indefinitely high. There is a characteristic concentration for a given enzyme and substrate known as the Michaelis constant (K_m) at which the rate is half the saturated rate. RuBisCO has two substrates, RuBP and carbon dioxide. The K_m for carbon dioxide is approximately $7 \mu M$, and the concentration of carbon dioxide in air-saturated water is only $10 \mu M$. Because of the need for carbon dioxide to diffuse through the stomata of the leaf, through the air spaces between the cells, and through the cell wall and cytoplasm to the chloroplasts, it is certain to be much lower than $10 \mu M$ at the point of reaction with RuBisCO, although the precise value is hard to estimate and may be different in different cells. The result is that the catalytic rate of RuBisCO is considerably less than the maximum expected from the saturated rate. In addition, we could expect that changes in the properties of RuBisCO would have major effects on the rate of photosynthesis: RuBisCO may be a rate-limiting step. However, an alternative view is that the principal rate determinant is the regeneration rate of $Ru-1,5-P_2$, and RuBisCO adjusts to match it.

RuBisCO, as an enzyme, is indeed subject to several kinds of modification, activation and control. It requires that carbon dioxide be bound to an A subunit, probably as a carbamate of an amino group, with Mg^{2+} ions.

Without this activation, the Michaelis constant for carbon dioxide is very much higher so that the enzyme appears to be inactive except in artificially high gas concentrations. An 'activase' enzyme is required to attach the carbon dioxide, and if the activator is missing (genetically) the plant can only fix carbon dioxide if the atmosphere is greatly enriched (Salvucci *et al.*, 1986). The activating carbon dioxide molecule is distinct from the substrate for the carboxylase reaction. The B subunit mediates the controlling effect of magnesium ions.

RuBisCO is inactive in the cell, in the dark. When RuBisCO is isolated, it must be reactivated by incubation with carbon dioxide and Mg^{2+}, if the activity is to be properly assayed. It is not certain whether the enzyme normally undergoes deactivation to any great extent in the chloroplast; it has been shown that the inactivation of RuBisCO that is observed during the hours of darkness is due primarily to the formation in the chloroplast of the material carboxyarabinitol 1-phosphate, which is a close analogue of the bound intermediate in the carbon dioxide fixation mechanism. This substance is known as the 'pre-dawn inhibitor'; the effect is rapidly removed because there is virtually a 1:1 molecular ratio of inhibitor to enzyme (Berry *et al.*, 1987). It is of major importance in the regulation of RuBisCO in *Phaseolus vulgaris*, for example, but apparently not in *Spinacea oleracea* (see the review by Woodrow and Berry, 1988).

RuBisCO requires Mg^{2+} ions for its reaction as well as for its activation, and it also has an alkaline pH optimum. Both of these conditions are improved by the operation of the light-reaction in the thylakoids, which accumulate H^+ ions, raising the stromal pH, and also release Mg^{2+} ions in exchange for H^+. The stromal pH rises from 7 to 8.5, and the concentration of Mg^{2+} from near zero to 5 mM.

These factors that affect the activity of RuBisCO can be regarded as (a) *regulators*, by means of which the cycle is balanced, and (b) *switches*, which result in the enzyme being totally inactive at night. The supply of ATP during darkness comes from respiration and since the reductive pentose cycle is an ATP-consuming system, a futile cycle could be envisaged in which the mitochondria and chloroplasts were recycling the carbon dioxide in an ATP-consuming perpetual motion. The oxygenase activity of RuBisCO, described later, would exacerbate the situation and lead to a nocturnal loss of carbon from the leaf.

Controls on other enzymes of the cycle are less obvious. Reference has already been made to the role of the thioredoxin system in maintaining the reduced state of the thiols necessary for the activity of FBPase; however it is by no means certain whether the importance of thioredoxin is that it reverses the continuous oxidation of the thiols by molecular oxygen (produced during photosynthesis), or whether the inactivation of FBPase by thiol oxidation at night (when thioredoxin cannot be reduced by ferredoxin) is important in switching off the reduction pentose cycle.

Although the individual enzymes have turnover times measured in milliseconds to seconds, seconds to minutes may elapse after the onset of illumination, before a leaf reaches a steady state. One of the inflections on the fluorescence induction curve (Figure 4.5) is considered to indicate the effect of the onset of carbon dioxide fixation on the level of ATP.

7.2.3 Photorespiration

Leaves of temperate plants (but not tropical grasses such as sugar-cane or maize) show a phenomenon known as the post-illumination carbon dioxide burst. When the illumination is interrupted the photosynthetic carbon dioxide uptake reverses, and for some 10–15 seconds the leaf produces carbon dioxide. The use of $^{14}CO_2$ shows that the post-illumination carbon dioxide burst is derived from metabolites that are rapidly labelled by photosynthesis, that is, newly fixed carbon is used in preference to the complement of respiratory substrates in the cell.

The use of $^{18}O_2$, which is non-radioactive and requires a mass-spectrometer for detection and measurement, showed that oxygen uptake occurs simultaneously with photosynthetic oxygen production. The effect was traced to the substance glycollic acid (see Zelitch, 1979 for a review), which is formed by all aerobically photosynthesising cells, although not to any great extent in the whole leaves of tropical grasses. Glycollate is formed from phosphoglycollate, which in turn is formed from RuBP and oxygen, catalysed by RuBisCO acting as an oxygenase (Bowes et al., 1971):

$$RuBP + O_2 \rightarrow 3PGA + phosphoglycollate$$

$$phosphoglycollate \rightarrow P_i + glycollate.$$

Both reactions take place at the same time depending on whether an oxygen or carbon dioxide molecule encounters RuBisCO loaded with RuBP. That is to say, oxygen and carbon dioxide compete; at relatively high levels of carbon dioxide the carboxylase reaction predominates, and at higher relative levels of oxygen the oxygenase reaction is the more rapid.

Under normal conditions, the carbon dioxide level of the air, and therefore of carbon dioxide in the photosynthesising tissue of a (temperate) leaf, is insufficient to saturate RuBisCO, and both the oxygenase and

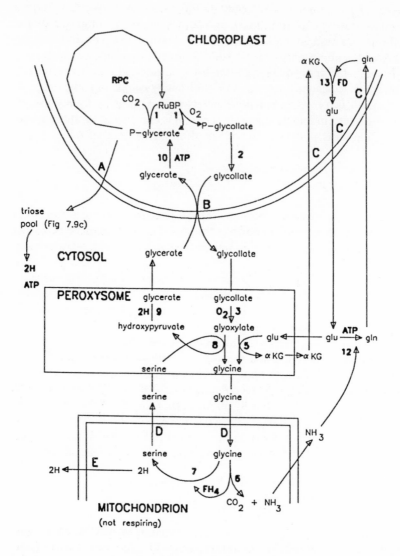

Figure 7.5 The C2 oxidative pathway of photorespiration.
Phosphoglycollate is generated in the chloroplast, for reasons that are a subject for discussion. It is scavenged in the multi-organelle pathway with the loss of $\frac{1}{4}$ of the carbon atoms as carbon dioxide. Arrowheads indicate the overall direction of flux, not irreversibility of reactions. Abbreviations: *FD*, ferredoxin (reduced); *FH₄*, the one-carbon pool carried by

carboxylase activities proceed. This accounts for the production of glycollate. The glycollate will possess the radiolabelling of C-1 and C-2 of RuBP. It also accounts to some extent for the uptake of oxygen. The output of carbon dioxide, however, does not come from the chloroplast, but is a whole-cell phenomenon. The glycollate leaves the chloroplast by means of a transporter which exchanges it for cytoplasmic glycerate. It passes through the cytoplasm and enters bodies known as peroxysomes (or glyoxysomes), where the enzyme glycollate oxidase, a flavoprotein, carries out the reaction:

$$\text{glycollate} + O_2 \rightarrow \text{glyoxylate} + H_2O_2.$$

The peroxide is decomposed by catalase:

$$2H_2O_2 \rightarrow O_2 + H_2O.$$

Glyoxylate is transaminated by glutamate or serine to form glycine. The glycine passes to the mitochondria, where the reactions take place catalysed by tetrahydrofolate-requiring enzymes:

$$\text{glycine} + FH_4 \rightarrow CO_2 + NH_3 + (CH_2)\text{---}FH_4$$

$$(CH_2)\text{---}FH_4 + NADH + H^+ \rightarrow (CH_3)\text{---}FH_4$$

$$(CH_3)\text{---}FH_4 + \text{glycine} \rightarrow \text{serine} + FH_4 + NAD^+$$

overall

$$2\,\text{glycine} + NADH + H^+ \rightarrow CO_2 + NH_3 + \text{serine} + NAD^+.$$

The ammonia, released in the above reactions, is re-fixed by means of glutamate synthase, which requires ATP, and the serine and glutamate return to the peroxysome where they transaminate the incoming glyoxylate. 2-oxoglurate returns from the peroxysome to the mitochondrion. Transamination from serine leaves hydroxypyruvate, which is reduced to glycerate in the peroxysome. Glycerate re-enters the chloroplast in exchange for outgoing glycollate, and is there phosphorylated to 3PGA. The pathway is set out in Figure 7.5.

Figure caption 7.5 (*Continued*)
the coenzyme tetrahydrofolate; *gln*, glutamine; *glu*, glutamate; 2H, NAD(P)H; αKG, α-ketoglutarate (2-oxoglutarate). The numbers shown refer to the name of the enzyme or reaction: (1) RuBisCO, (2) phosphoglycollate phosphatase, (3) glycollate oxidase, (5) glutamate/alanine:glyoxylate aminotransferase, (6) glycine decarboxylase complex, (7) serine hydroxymethyltransferase, (8) serine:glyoxylate aminotransferase, (9) NADH:hydroxypyruvate reductase, (10) glycerate kinase, (12) glutamine synthetase and (13) glutamate synthase. Translocators of intermediates of the pathways are indicated by the letters: (A) phosphate translocator, (B) glycolate/D-glycerate translocator, (C) chloroplast dicarboxylate translocator(s), (D) mitochondrial amino acid translocator(s) and (E) mitochondrial dicarboxylate translocators. This scheme was largely elucidated by N.E. Tolbert. From Husic *et al.* (1987).

The photorespiration pathway in cyanobacteria (prokaryotic; there are no glyoxysomes) proceeds differently: two molecules of glyoxylate condense to give carbon dioxide and tartronic semialdehyde:

$$2\,CHO \cdot COOH \xrightarrow[\text{glyoxylate carboligase}]{} CO_2 + CHO \cdot CHOH \cdot COOH$$

$$CHO \cdot CHOH \cdot COOH + (2H) \xrightarrow[\substack{\text{tartronic semialdehyde} \\ \text{reductase}}]{} CH_2OH \cdot CHOH \cdot COOH$$

What is the significance of photorespiration? Only two suggestions seem reasonable: first, that the cycle as drawn is misleading and that its real significance is the production of amino acids such as serine and its derivatives from photosynthetic products, or alternatively that the cycle is a scavenging system to recover as much as possible (that is three-quarters) of the carbon lost by the oxygenase activity of RuBisCO.

Is there any value in the oxygenase activity? It is found universally, even in the RuBisCO from photosynthetic bacteria that never encounter oxygen. Both activities share the same active centre on the enzyme, and the view is taking hold that oxygenase activity is an inevitable concomitant of carboxylase activity. In activating RuBP for attack by carbon dioxide, RuBisCO renders it vulnerable to attack by oxygen.

If, as has been suggested in plants such as wheat, about half the carbon dioxide initially fixed by the reductive pentose cycle is subsequently lost by the oxygenase activity of RuBisCO and the subsequent photorespiration, then there could be immense agricultural and economic gain to be realised if a differential inhibitor could be found that would stop the oxygenase, while permitting the carboxylase activity to continue. Nothing has so far appeared in sight. Is this idea in fact valid? The following consideration of photoinhibition may suggest that it is not.

Photoinhibition. Isolated and broken chloroplasts are rapidly inactivated if they are illuminated in the absence of an electron acceptor. The main site of the damage is the D1 protein in photosystem II. Damage also appears to take place even in intact leaves, and the D1 protein is continually being resynthesised (see Schuster *et al.*, 1988, for a review). The D1 and D2 proteins, according to the model of photosystem II previously described, are subject to a very considerable electrical stress insofar as they carry the positive charge of P680 or the tyrosine radical Z, and the negative charge in phaeophytin, Q_A or Q_B, separated by only a few tens of nanometres. It may not be too great a simplification to suppose that if the rate of electron transport is diminished by the lack of electron acceptors, then the charges and the associated electrical stress remain on the D1 protein for longer, and the rate of damage increases beyond the capacity of the chloroplast to repair it.

There is a relatively long diffusion path for carbon dioxide from the air via the stomata (or through the cuticle where it is thin enough), through the air spaces between the cells, and, in solution, through the cell wall and cytoplasm to the chloroplasts, and through the envelope membranes to the stroma and the RuBisCO enzyme. The diffusion path is liable to be restricted by closing of the stomata in response to water stress, high temperature, high light levels or diurnal rhythm. The concentration of carbon dioxide in air-saturated water is close to the K_m of RuBisCO (the concentration at which the rate of reaction is half the rate when the concentration is saturating). Therefore, the situation must frequently occur when the natural electron acceptor of the light-reactions in the chloroplast, carbon dioxide, is scarce and restricting the rate of electron transport, rendering the plant susceptible to dangerous photoinhibition. The oxygenase activity of RuBisCO may well be essential to maintain the level of carbon dioxide at the compensation point, so that the rate of electron transport is not greatly restricted, and damage to the D_1 protein is minimised. The net loss of carbon (the plant may lose carbon dioxide to the atmosphere) during the midday stomatal closure may be a trivial price to pay for survival.

There is an abundance of the enzyme carbonic anhydrase in photosynthetic tissues, which is perhaps surprising since carbon dioxide enters and is consumed as such. It may be that there are active transport mechanisms in membranes that allows the cell to force the accumulation of bicarbonate ions; this is particularly likely in aquatic plants.

Phosphorylation of thylakoid proteins: a defence? A number of proteins in the thylakoid undergo phosphorylation and dephosphorylation. The best characterised is the antenna complex LHCIIβ. The complex acquires phosphate groups from ATP by means of a protein-kinase located in the thylakoid membrane, probably attached to the cytochrome *bf* complex. The activity of the kinase is governed by the degree of reduction of plastoquinone; when PQ is mostly in the form PQH_2 the kinase is most active, and is almost inactive when the quinone is in the oxidised form. There is a phosphatase enzyme that hydrolyses the phosphate groups from

LHCII, which does not appear to be controlled in any way:

$$LHCII + ATP \xrightarrow{\text{kinase}} LHCII\text{-}P + ADP$$

$$LHCII\text{-}P + H_2O \xrightarrow{\text{phosphatase}} LHCII + P_i.$$

The result is that when plastoquinone is mostly reduced, the kinase outweights the phosphatase and LHCII becomes mostly phosphorylated. When the quinone is mostly oxidised, LHCII loses its phosphate.

Note that PSII can only perform electron transport when PQ is available to act as an electron acceptor. If there is no PQ available the bound semiquinone Q_A and the PQH_2 at Q_B will have an indefinitely long lifetime during which they are vulnerable to attack, by for example oxygen. The point of the phosphorylation of LHCII is that it acquires extra negative charge, which repels it from PSII. It is possible but not proved that LHCII when phosphorylated can associate with PSI. Hence, light energy is diverted away from PSII and into PSI, the rate of oxidation of PQH_2 increases, and the rate of arrival of electrons from PSII diminishes: the system comes into balance.

The changes in the phosphorylation state of LHCII take several seconds, up to a minute, to complete, and the fluorescence characteristics of the chloroplast change from 'State I' (the normal state) to 'State II' (where LHCII is disconnected from PSII).

It is also suggested that the tight stacking of thylakoids to form grana in higher-plant chloroplasts is relaxed in State II, and this will also alter the light-absorption characteristics.

Other proteins are also phosphorylated, such as the 10 kDa polypeptide in PSII. The rationale may or may not be the same as for LHCII.

7.2.4   C4 photosynthesis

Tropical grasses are notable for avoiding photorespiration. This is a property of the whole leaf. The photosynthesising cells surround the veins of the leaf and are of two types, bundle-sheath, that form a tight single layer, with no air spaces, round the vein, and surrounding the bundle-sheath some larger cells with air-spaces known as mesophyll (a diagram is given in Figure 7.6). This is known as Kranz (wreath) anatomy (there are various types). Together with the restriction of stomata and the thick cuticle on the leaf epidermes it represents an adaptation to semi-arid environments. There is a difference between the forms of the chloroplasts in the bundle-sheath

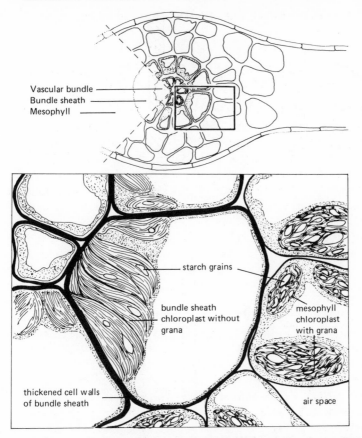

Figure 7.6. Diagram of Kranz anatomy. A cross-section of a leaf of crabgrass *Digitaria sanguinalis* is shown. Drawn from micrographs of Black and Mollenhauer (1971).

(BS) and mesophyll (MES) cells. It may be a matter of size (BS being larger) or of grana content (BS being deficient in grana in for example maize).

It is possible to peel off the epidermes and to digest the exposed cell walls of the leaf interior with cellulase and pectinase enzymes, and by this means the cell types can be isolated. MES cells from maize are found to produce oxygen in light and take up carbon dioxide into C4 organic acids, chiefly malate and aspartate (in equilibrium with a small concentration of oxaloacetate). They do not contain RuBisCO. BS cells from maize do not produce oxygen (PSII is inactive) but they do contain RuBisCO and can fix

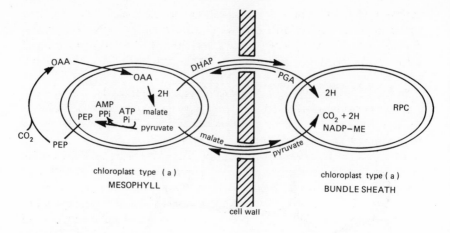

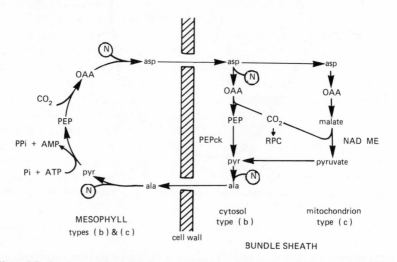

Figure 7.7. C4 photosynthesis. The mesophyll cells (on the left) carry out the conversion of pyruvate to phosphoenolpyruvate (PEP) in the chloroplasts, and the carboxylation of PEP to oxaloacetic acid (OAA) in the cytoplasm. In type (a), the OAA returns to the chloroplast where it is reduced to malate, which travels to the bundle-sheath cell. It is decarboxylated by the NADP-linked malic enzyme. Type (a) chloroplasts also pass reducing equivalents to the bundle-sheath by means of the DHAP/PGA shuttle (see Figure 7.9). Mesophyll chloroplasts of types (b) and (c) transaminate OAA to aspartate (asp) in the cytosol, and pass it to the bundle-sheath in exchange for alanine (ala). Bundle-sheath cells of type (b) decarboxylate OAA by means of the cytosolic PEP-carboxykinase (PEPCK) whereas those of type (c) convert OAA to malate in the mitochondria, and decarboxylate it by means of the NAD-linked malic enzyme. Plasmodesmata are indicated in the cell wall.

carbon dioxide by means of the reductive pentose cycle, obtaining their reducing power from malate. Not all C4 plants are the same; some retain PSII activity in their BS chloroplasts, and they differ in the relative use of malate and aspartate as the means of communication between the two cell types (Figure 7.7).

In MES cells the formation of phosphoenolpyruvate (PEP) by means of a single enzyme is unusual; the very high phosphate-transfer potential of PEP requires that two phosphate groups be lost from ATP, via the temporary phosphorylation of P_i. Pyrophosphate is very rapidly hydrolysed in the cell, and reactions in which pyrophosphate is produced are rendered irreversible by this means. Note that this reaction is similar to some extent with that found in the green bacteria associated with the reductive citrate cycle.

Carbon dioxide is liberated in the BS cells by the various alternative reactions shown in Figure 7.7, and the consequence of the expenditure of ATP in the MES cell is that even after allowing for the diffusion of the C4 acids from MES to BS cell, the carbon dioxide is released at a much higher concentration than would have been available from air alone. This is remarkable because the same adaptations that serve to retain water vapour within the leaf also restrict the entry of carbon dioxide. The MES cells therefore serve as a very effective system for pumping carbon dioxide from the air spaces of the leaf into the stroma of the BS chloroplasts. The carbon dioxide concentration is raised to levels at which RuBisCO is more nearly saturated, and the carboxylase reaction competes effectively with the oxygenase. The absence of PSII and hence of oxygen production in BS cells of the maize type further improves the CO_2/O_2 balance. The effect is that the C4 group(s) of plants are very efficient users of solar energy in hot climates with bright sunlight. Sugar-cane, one of the most efficient plants, is said to conserve more than 1% of the photosynthetically-active solar radiation. When light and temperatures are at lower levels, that is in temperate climates, C4 plants become progressively less successful, because of the need to assimilate each carbon dioxide molecule twice; full sunlight is a more rare occurrence. Additionally, the lower attainable leaf-temperatures slow down diffusion and reaction rates, and the more complex C4 pathway is slowed down more in comparison with the temperate (C3) plants.

Crassulacean acid metabolism. The succulent plants chiefly including the Crassulaceae use the same reactions to accumulate C4 acids during the night, so that by dawn the vacuoles are perceptibly acid. These plants are often found in deserts; during the night the stomata can be opened to take in

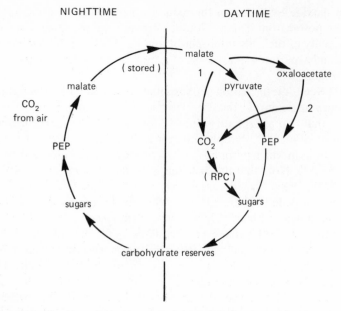

Figure 7.8. Crassulacean acid metabolism (CAM). (a) Night-time fixation of carbon dioxide into C4 acids. (b) Daytime release of carbon dioxide by means of either (1) PEP-carboxykinase or (2) NADP-linked malic enzyme. Carbon dioxide is fixed by the normal operation of the reductive pentose cycle (RPC).

carbon dioxide without severe water loss, and during the day the stomata are closed and the carbon dioxide released and retained within the structure. This amounts to a carbon dioxide storage system, or a 'temporal' pump. The overall scheme (Figure 7.8) is similar to the C4 system (Figure 7.7), except that the C3 acids are produced during the night from polysaccharide, probably starch, and returned during the day when the C4 acids are broken down. It is important that both C4 and crassulacean acid metabolism (CAM) both ultimately depend on the reductive pentose cycle for the permanent fixation of carbon dioxide.

Unsolved problems are the long-term controls. CAM plants have a 'biological clock' which switches the two halves of the C4 cycle to correspond with day and night. This clock can be re-set when the photoperiod is experimentally altered. The C4 habit in some plants may be lost altogether if conditions of water stress are removed.

7.2.5 Chloroplasts and respiration

Chloroplasts do not respire; photorespiration depends on the cooperation of chloroplast, peroxysome and mitochondrion. Normal respiration, carried on by the mitochondria, provides ATP for the cell during the hours of darkness, but there is evidence that it is switched off in the light. This is shown by the Kok effect: B. Kok measured the quantum requirement for photosynthesis, which is the number of quanta of light that are absorbed for every molecule of carbon dioxide consumed or oxygen produced. The value had been shown by R. Emerson to be 9, which was interpreted as 8 plus inefficiencies in the system. The value of 8 is consistent with two light-driven photoreactions and a four-electron process. Kok's measurements showed that in dim light, when the rate of (mitochondrial) respiration outweighed photosynthesis, the quantum requirement for the apparent carbon dioxide output (that is, the diminution in the actual carbon dioxide uptake) was half the expected value, that is, 4. The explanation now accepted is that as the chloroplast activity increases in the increasing illumination, the mito-chondrial activity diminishes. The mechanism by which the chloroplast influences the mitochondrion is not clear: it has been suggested that the chloroplast takes up inorganic phosphate, making it unavailable to the mitochondrion.

A stranger topic is 'chlororespiration'. It has been suggested that some chloroplasts can oxidise the coenzyme NADH. In fact chloroplast DNA carries some genes that are very similar to those in mitochondria, coding for iron–sulphur proteins in complex I, the NADH:ubiquinone reductase (Ohyama et al., 1988; Shinozaki et al., 1988). It is of course possible that these are 'fossil' genes, and are not expressed, but it is an interesting possibility that these chloroplasts are able to produce something resem-bling complex I. The value of such a complex would be not so much as an NADH-oxidising system, but as a means of producing NADH by reverse electron transport (section 6.6).

7.3 Chloroplast–cytoplasm transport

The two membranes that form the chloroplast envelope contain carotene, but no chlorophyll. There are different populations of proteins, which enables the experimenter to separate inner and outer envelopes on the basis of the greater density of the former. (See Douce and Joyard, 1979 for a reivew).

The envelope is important in the synthesis of the lipids of the chloroplast thylakoid. These are predominantly galactolipids, which are formed in the inner envelope. The outer envelope contains a higher percentage of phospholipids, which are the predominant lipid of other cell membranes. The fatty acids contained in galactolipids are of two classes, known as prokaryotic and eukaryotic, and occur in thylakoids in different proportions in different plants. The inner membrane is the site of synthesis of the prokaryotic group, which, simplifying somewhat, is characterised by a 16-carbon fatty acid at position-2 of glycerol, as opposed to an 18-C acid at position-2 in the eukaryotic series. Moreover, while both families are rich in polyunsaturated patty acids, the prokaryotic group tend to have more 16:3 (hexadecatrienoic acid) and the eukaryotic group more 18:3 (linolenic acid). The fatty acids themselves are synthesised exclusively in the chloroplast stroma via acetyl CoA and acetyl CoA carboxylase, an enzyme which has been identified as the target of a potent new series of graminaceous herbicides (Harwood, 1988).

Reference has been made to a transporter in the chloroplast envelope that exchanged glycollate and glycerate. Such transporters are found in the inner envelope which is the effective barrier for small molecules. The outer envelope is pierced by porin, a multi-subunit enzyme that forms a pore. In chloroplasts the pore can pass molecules up to 10 kDa in mass; similar but smaller porins are known in the bacterial periplasmic membrane and the mitochondrial outer envelope.

It is established that ATP and other nucleotides can penetrate the envelope, but slowly, sufficient for growth but not sufficient to account for the ability of a chloroplast to provide ATP for use in the cell. That this does indeed happen is shown by the cell carrying out ATP-dependent functions in the absence of oxygen (hence without respiration) and with a requirement for light; if the light is of sufficiently long-wavelength (above 700 nm) to avoid exciting PSII the processes still work and it is assumed that ATP is produced by the cyclic electron-transport system. Examples of such cellular processes are the uptake of K^+ ions or glucose by algal cells, and protoplasmic streaming in many leaves.

The 'phosphate-transporter' can exchange any two of P_i, GAP, DHAP, 3PGA and probably PEP. The total phosphate content of the chloroplast is thus kept constant. (Intact chloroplasts suspended in phosphate buffer are liable to lose their organic substrates by exchange, with consequent inhibition; it can be prevented with the use of pyrophosphate buffer.) Another transporter (without necessary exchange) operates on dicarboxylic acids such as malate, aspartate, glutamate and oxoglutarate. Figure 7.9 shows how transporters can work together so that the chloroplast is able to export to the cytoplasm any combination of NADH, ATP and

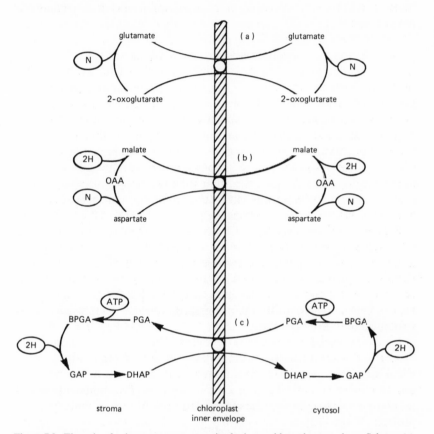

Figure 7.9. The role of substrate-transporters in the inner chloroplast envelope. Scheme (a) represents the dicarboxylate transporter in a shuttle which transfers amino-nitrogen across the chloroplast envelope. The shuttle can work in both directions, either outwards (the chloroplast is a net exporter of glutamate) or in conjunction with other shuttles (see Figures 7.7 and 7.5). N represents the amino group carried by the coenzyme pyridoxal phosphate. Scheme (b) also represents the dicarboxylate transporter, carrying reducing equivalents (2H, on NADP inside the chloroplast, NAD in the cytosol), and amino groups, in opposite directions. Scheme (c) shows the phosphate translocator exchanging DHAP and PGA and thus effectively transporting ATP and (2H) out of the chloroplast. If all three (a, b and c) function simultaneously, ATP is transported with no other gains or losses.

aminonitrogen. Oxaloacetate is not a transportable substrate, partly because it has a very low concentration in equilibrium with malate and NAD(P)H.

7.3.1   The products of photosynthesis: starch and sucrose

The chloroplast can, as shown in Figure 7.9, take in carbon dioxide, and export triose phosphate (as dihydroxyacetone phosphate) in exchange for inorganic phosphate. Triose phosphate is in equilibrium with fructose-1,6-bisphosphate, and the latter compound is the most abundant in the equilibrium mixture. When a leaf is illuminated after a period of darkness, the concentrations of all these metabolites rise, both in the cytosol and in the chloroplast. It will be recalled that the reduction of phosphoglyceric acid to triose phosphate was close to equilibrium, and hence the concentration of PGA rises simultaneously with GAP, DHAP and F16BP.

F16BP can only react in two directions: it can re-form the triose phosphates (which are in equilibrium with F16BP) or it can hydrolyse to form fructose-6-phosphate. The enzyme that catalyses this second reaction, F16BPase, is strongly regulated by its inhibition by another fructose compound, fructose-2,6-bisphosphate (F26BP). F26BP is formed from F6P by specific kinase (ATP-requiring phosphotransferase) which itself is strongly inhibited by DHAP and activated by F6P and P_i. F26BP is broken down by a phosphatase back to F6P and P_i. The phosphatase is inhibited by F6P and P_i. (When we say that an enzyme is inhibited by its products, we mean more than just the increase in rate of the reverse reaction; we mean that the enzyme is altered and becomes less active.) The combined effect of these controls provides an exquisite regulation of the flow of photosynthetic product (Figure 7.10).

In the dark, the cell respires, and there is a flow of carbon from the starch grains to triose phosphate which passes out of the chloroplast and down the glycolysis pathway to pyruvate, finally being respired by the mitochondria. The hydrolysis of F16BP in the cytosol is not required, and accordingly the F16BPase is inhibited by the high level of F26BP caused by the activation of the F6P 2-kinase and the inhibition of the F26BPase (by high cytosolic levels of P_i in both cases).

When illumination commences, the chloroplast takes up P_i and exports DHAP. The formation of F26BP slows and its breakdown increases: the level of F26BP falls, releasing the inhibition of F16BPase. F6P is then formed, which equilibrates with glucose-6-phosphate (G6P). G6P equilibrates with glucose-1-phosphate (G1P). G1P is activated by uridine

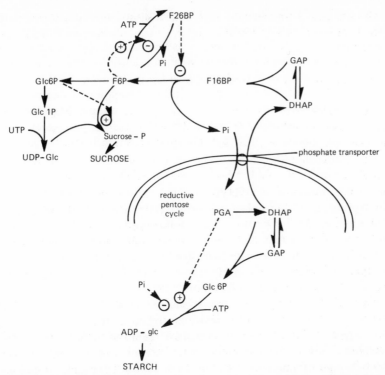

Figure 7.10. Formation of products, starch and sucrose, and partitioning. The total phosphate inside the chloroplast is kept more or less constant, so that if the organic phosphates increase, P_i falls. The formation of ADP-glucose, on the route to starch, is activated by PGA and inhibited by P_i. Starch formation is therefore secondary to sucrose formation outside the chloroplast, controlled at the fructose-1,6-bisphosphate (F16BP) 1-phosphatase reaction by fructose-2,6-bisphosphate (F26BP), itself regulated by the cytosolic levels of fructose-6-phosphate (F6P) and P_i. Symbols: $X\cdots\to +$, activation by metabolite X; $X\cdots\to -$, inhibition by X.

triphosphate (UTP):

$$UTP + G1P \xrightarrow{\text{pyrophosphorylase}} UDPG + PP_i$$

$$PP_i \xrightarrow{\text{pyrophosphatase}} 2P_i.$$

The uridine diphosphate part of UDPG is displaced by F6P (on the enzyme sucrose phosphate synthase) yielding sucrose phosphate, which is hydrolysed by sucrose phosphatase to sucrose and P_i. Sucrose is stored in the vacuole of the leaf cell, and translocated out of the leaf into the phloem of the vascular bundles.

The rate of translocation of sucrose is limited. The synthesis of sucrose is controlled so as to limit the cytosolic concentration, by means of an increase in the level of F26BP (mediated by an unknown mechanism).

Meanwhile, inside the chloroplast, once the formation of sucrose in the cytosol has reached its maximum rate, the concentration of PGA rises and P_i falls. This activates an enzyme in the chloroplast stroma which forms ADP-glucose (ADPG) by a pyrophosphorylase reaction analogous to UDPG above:

$$ATP + G1P \rightarrow ADPG + PP_i.$$

The glucose moiety of the ADPG is added to the starch molecules in the starch grain in the chloroplast, releasing ADP. Therefore, the events in the flow of the product of photosynthesis following the onset of illumination are (a) there are rapid adjustments in the levels of P_i and DHAP; (b) sucrose synthesis commences, and (c) sucrose export and starch synthesis commence. These events, collectively known as carbon partitioning, take a minute or more to accomplish (Figure 7.10).

The major role of P_i as a metabolic regulator should be noted; in mammalian liver there are similar interrelationships between P_i and Fru26BP, but in addition the enzymes are regulated by being themselves phosphorylated by protein kinases, under the control of hormones. Protein phosphorylation in plant cells certainly takes place, but its importance and control is a new and largely unknown field.

Starch grains are dissolved during the hours of darkness, probably by the enzyme phosphorylase:

$$(Glu)_n + P_i \rightarrow (glu)_{n-1} + glu1P.$$

It is not known how starch breakdown is controlled. Glu1P is metabolised to triosephosphate for export from the chloroplast in the form of DHAP. An important enzyme in this pathway is phosphofructokinase (PFK) which converts Fru6P to Fru16BP. In animal tissues PFK is known to be a major point of control in the glycolysis pathway. It would be expected that the cell would contrive not to have PFK active simultaneously with F16BPase, since a 'futile cycle' would be set up in which the only effect would be the destruction of ATP. However, in such tissues the activity of PFK is much greater than that of F16BPase (in the glucogenesis pathway), and it may be that simultaneously-active enzymes may provide a means of regulating glycolysis; instead of 'futile cycle' call it a 'substrate cycle'. In the chloroplast FBPase is very active, and PFK is almost certainly regulated to prevent a futile cycle, probably by inhibition at high levels of ATP.

The ATP required for the operation of PFK, in the chloroplast in the dark, must come from the cytosol by the reverse operation of the shuttle system described in Figure 7.9.

7.3.2 Guard cells

A leaf is a complicated structure requiring to present an extended area to light, but to be thick enough to retain mechanical stiffness, yet without restricting the supply of carbon dioxide to its photosynthetic cells. This is achieved by the sandwich structure: adaxial epidermis–palisade tissue–spongy mesophyll–abaxial epidermis. Carbon dioxide and oxygen diffuse rapidly in the air spaces surrounding the mesophyll cells, and only a relatively short distance has to be travelled by means of diffusion in solution through cell walls and cytoplasm.

Higher plants have a problem of water conservation. The two epidermes have an outer waxy cuticle that restricts water loss but in order to allow the necessary gas exchange there are holes known as stomata (singular: stoma) that are bordered by pairs of cells known as guard cells. These have a different shape from other epidermal cells, and unlike them they contain chloroplasts. Changes in turgor pressure affect the shape of the guard cells by virtue of the uneven thickness of the walls; in general, loss of turgor causes the aperture to close. Stomatal behaviour varies widely from plant to plant, but it is commonly found that stomata tend to close in darkness and in very intense light. They may also close at midday in a circadian rhythm. The chloroplasts of guard cells are different from those in the palisade or spongy mesophyll; they produce C4 acids, and the reductive pentose cycle is absent. Changes in turgor pressure result from uptake or loss of water at the plasmalemma of the cell, under the influence of a blue-light receptor.

Adaptations to environmental aridity include thickening of the cuticle, restricting the stomata to one surface of the leaf, and protecting the surface by hairs or by rolling the leaf. All these changes are observed to accompany a slowing of the rate of growth compared to typical temperate plants. Water conservation carries a penalty in restricting gas exchange, hence the value of the C4 habit in the tropical grasses, when the light-levels and temperatures are sufficiently high.

7.3.3 Carbon dioxide accumulation

Evidence is accumulating (see Kelly, 1984) that many algae and some higher plant cells are able to trap and store carbon dioxide by a process in which carbonic acid is formed by means of the enzyme carbonic anhydrase, and then combines (weakly) with protein (on amino groups) in the form of carbamate. This mechanism is known to exist in vertebrate red blood cells for the transport of carbon dioxide attached to haemoglobin. This is not concentration in the sense that applied to C4 photosynthesis, where owing to the use of ATP the concentration of carbon dioxide was increased. Carbonic anhydrase does not use ATP. Nevertheless a pool of trapped carbon dioxide in the cell provides a means of keeping photosynthesis going when there is a local deficiency in the gas or when stomata are closed.

7.4 Molecular cell biology

Metabolic adjustments occur in time-scales of a few minutes, which allow for raising or lowering the levels of individual metabolites and setting up diffusion gradients. Changes can also be made to the absolute quantities of enzymes present, and this requires protein synthesis in a time-scale of hours.

7.4.1 *Genes exist in the cell nucleus and in organelles*

Essentially, protein synthesis follows the pattern whereby a gene is activated by some stimulus. Genes are recognisable segments of DNA, which is found both in the chloroplast and in the chromosomes of the nucleus. DNA is a very long linear sequence of deoxynucleotides; the genes are sections of this long sequence, and are believed to be activated by the attachment of proteins to specific sequences at the beginning of the gene. The activated gene is transcribed into RNA. RNA is a sequence of (ribo)nucleotides, corresponding by a rule to the particular sequence of deoxynucleotides in the gene. RNA molecules are long, but represent only a small fraction of the DNA. Often RNA molecules are reduced in length or divided up by nuclease enzymes before the RNA can be used. Some RNA forms part of ribosomes (rRNA), and others are the transfer RNA (tRNA) that specifically carry amino acids to the ribosome. Genes for proteins are transcribed into messenger RNA (mRNA) which serves as the template on which the ribosome assembles the polypeptide. Many mRNAs are produced by 'splicing' a primary RNA transcript, cutting out sequences of apparently useless material known as introns. Introns are generally regarded as a feature of nuclear genes but are found in organelles also (Figure 7.11).

The ribosome travels along the mRNA assembling amino acids as directed by the sequence of nucleotide codons until a stop codon is encountered. The polypeptide is detached, and is frequently a precursor of the target protein, requiring proteolytic pruning of its termini.

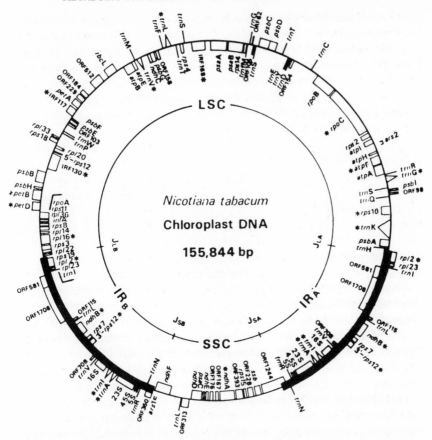

Figure 7.11. The chloroplast genome: the circular gene map. Genes are labelled conventionally (see Crouse *et al.*, 1985); labels on the outside are on one strand ('A') and are transcribed counterclockwise while those shown on the inside of the circle are on the 'B' strand and are transcribed clockwise. Asterisks indicate genes containing introns. LSC, large single-copy region; SSC, small single-copy region; IR$_A$ and IR$_B$, the two repeated sequences (marked on the circle with a heavy line); J$_{LA}$ (etc), junction between LSC and IR$_A$ regions; ORF228 (etc), an open reading frame that could code for a protein of 228 amino acids. From Shinozaki *et al.* (1988).

Most of the proteins found in the chloroplast are coded on nuclear DNA, and the mRNA transcripts, after removal of introns, are translated into protein by means of the 80S cytoplasmic ribosomes. The reagent cycloheximide is a specific inhibitor of 80S ribosomes, and proteins that are synthesised in the cytoplasm do not appear in the presence of cycloheximide. The newly synthesised proteins have to pass through the double envelope of the chloroplast to reach the chloroplast stroma, and through the thylakoid membrane if they are destined for the thylakoid lumen. In virtually all cases the first-formed protein is larger than the final, mature, form and possesses a 'leader sequence' at the N-terminal end. Sometimes parts of the leader sequence can be recognised as a 'signal peptide' that may engage with a specific transporter. Other sequences of interest are 'transit peptides' that are thought to be membrane soluble. The mental picture is of the protein initially burrowing through the membrane, and of the transit peptide being cleaved off by a specific proteinase so as to render the protein more hydrophilic and thus complete its passage. Mental images fail to make much sense of the organising process for intrinsic membrane proteins that have several membrane-spanning segments, connected by hydrophilic loops. In some cases proteins have precursors with trailing sequences, and these are likely to be involved in their addressing and despatch, in ways yet to be discovered. A general term for the parts of a precursor protein that are cut off in the maturation process, and are thought to be involved in addressing and despatch, is 'topogenic sequences'.

The chloroplast itself contains DNA (cpDNA), which is recognisably different from the nuclear DNA in its density, containing usually a higher proportion of the bases G and C than A and T. Whereas the nuclear DNA is packaged into dense coils round histone particles, forming nucleosomes, there is much less protein associated with chloroplast DNA. The most striking difference however is that the chloroplast DNA has no ends, and is often described as circular, with a circumference of some 45 μm. In this respect chloroplast DNA closely resembles that of bacteria, although in size cpDNA is smaller; cpDNAs contain some 120 000 (*Marchantia*) or 150 000 base pairs (*Nicotiana*) compared with 2 million in *E. coli* or 2000 million in man. There are however up to 100 copies of the chloroplast DNA in each chloroplast. The chloroplast genome (Figure 7.11) codes for all the rRNA and many of the proteins that are required for making the characteristic 70S ribosomes, and it also has code for at least one of the tRNAs (symbol: *trn* + the single letter code for the amino acid) that are required for the transport of each of the twenty amino acids to the ribosome. It is expected that complement of tRNA is sufficient, even though some codons (triplets of bases) are not represented.

The list of identified genes in the *Nicotiana* genome appears in Table 7.1. A similar map and list has been obtained for the liverwort *Marchantia polymorpha* by Ohyama (see Ohyama *et al.*, 1988) and the differences are slight. In fact, judging from the (incomplete) structure of many chloroplast genomes, the main differences involve duplication of some sequences, or the

Table 7.1 Genes identified in the chloroplast genome of *Nicotiana tabacum*.

Genes	Gene products
RNA genes	
rRNA	rRNA (16S, 23S, 4.5S, 5S)
trn	tRNA (30 species)
Photosynthetic apparatus	
*rbc*L	RuBisCO large subunit
*atp*A,B,E,F,H,I	ATP synthase: CF_1 subunits α, β, e
	CF_0 subunits I, III, IV
*psa*A,B,C	PSI: Ia, Ib, 9 kDa subunits
*psb*A,B,C,D,E,F,G,H,I	PSII: D_1, CP47, CP43, D_2, 9 and 4 kDa units of cyt.
	b-559, G-protein, 10 kDa PP, I-protein
*pet*A,B,D	$b_6 f$ cplx: cyt. *f*, cyt. b_6, unit IV
Respiration or reverse electron transport	
*ndh*A,B,C,D,E,F	NADH dehydrogenase cplx. (5 subunits)
Gene expression	
rps(n)	30S ribosome subunit (proteins: n = 2, 3, 4, 7, 8, 11, 12,
	14, 15, 16, 18, 19)
rpl(n)	50S ribosome subunit (proteins: n = 2, 14, 16, 20, 22, 23,
	33, 36)
*rpo*A,B,C	RNA polymerase α, β, β' subunits
inf A,	Initiation factor I
ssb	ssDNA binding protein

PP, phosphoprotein. The G- and I-proteins were identified in PSII by their reaction with antibodies raised against synthetic oligopeptides made according to the DNA sequences now known as *psb*G and I. Source: Table 1 of Shinozaki *et al.* (1988).

omission of one of the inverted repeats. Ohyama lists 33 unidentified open reading frames (ORF) varying in size from 29 equivalent amino acids to 2136. Most are found in *Nicotiana* also, although so far extensive searching of genetic databases has failed to match these ORF to any known genes elsewhere.

7.4.2 The control of chloroplast synthesis

Nearly all the proteins of the chloroplast are encoded on genes in the cell nucleus. The protein products of chloroplast genes combine with the transported and processed nuclear gene products and form multi-subunit proteins. It is observed that when through genetic mutation or experimental inhibition the nuclear components are not produced, the corresponding chloroplast components are suppressed also. Ellis (1977) regarded this as the 'cytoplasmic control principle' and stressed that the inverse (suppression of nuclear-controlled synthesis by the lack of organelle products) did not occur. In fact nuclear-derived subunits do not accumulate in the inhibited chloroplast because there is a proteolytic system that destroys them, but their continuing synthesis is revealed by labelling with ^{35}-S-methionine. In this way Bennett (1981) showed that the major polypeptide of LHCII is broken down in plants which fail to produce chlorophyll b.

The mechanism of cytoplasmic control is exerted both at the transcription and translational phases. The field is 'active', which is a way of saying that there is no clear answer. However, the chloroplast codes for some subunits of its RNA polymerase. The promoters of plastid genes belong to the general prokaryote type. Once transcription was activated, the polymerase level could increase autocatalytically. Translational control is even less clear, but the chloroplast codes for the RNA-content of its ribosomes, and a number of ribosomal proteins, but there must be additional factors to explain the relatively sudden translation of existing mRNA, when algal cells are illuminated. Mullet (1988) has reviewed this area.

Photogenes. Protein synthesis is under several types of control, of which two are of special interest to photosynthesis. Developing tissue does not, in most plants, become green in the dark, and the resulting shoot is pale yellow, long and weak, with small leaves. This is known as *etiolation*. The cells contain etioplasts, which are clearly a form of chloroplast in which there is no chlorophyll, and no lamellar thylakoid structure. The membranes in the etioplasts instead form a three-dimensional tubular array known as the prolamellar body. In the electron microscope the prolamellar body has a crystalline appearance. The tubular form is probably taken up naturally, because the protein complement is very low, and protein complexes are important in maintaining the form of thylakoids. It must be noted that etioplasts are a response to growth in darkness; plants grown normally in daylight do not form etioplasts, but chloroplasts first appear as granumsized primordia.

Photomorphogenesis: protochlorophyllide. In the etioplast, synthesis of chlorophyll stops at protochlorophyllide, bound to a protein; the complex is protochlorophyll holochrome. The absorption of light by protochlorophyllide triggers two things: the reduction of protochlorophyllide to chlorophyllide and then chlorophyll, and the switching-on of the production of the intrinsic thylakoid membrane proteins. The accumulation of protein corrects the lipid–protein balance so that the tubular structure becomes the lamellar structure of mature thylakoids. These proteins are said to be under the control of photogenes acting on the chloroplast genome.

Photomorphogenesis: phytochrome. A second set of photogenes acts on the nucleus, and is likely to be controlled by the pigment phytochrome. Phytochrome is a biliprotein which undergoes conformational changes when it absorbs light, and has two spectral forms, the red-absorbing form, P_R (670 nm) and the far-red form, P_{FR} (730 nm). Phytochrome is involved in many physiological effects which depend on the colour-balance of light, or on daylength. The colour balance may be altered by factors such as the depth of soil overlaying a seed, or with more relevance to photosynthesis, by the presence or absence of an overhanging leaf. Thus, actions of phytochrome in controlling seed germination and the growth and form of a plant may be due in part to its ability to act as a chlorophyll detector.

Such evidence as is available for the mode of action of phytochrome provokes the thought of analogy with the action of certain mammalian hormones, insofar as the primary stimulus (absorption by P_{FR}, or a hormone combining with a receptor in the cell membrane) generates a second messenger that affects either transcription of DNA or translation of mRNA. The significance of phytochrome in the development of leafy plants is that leaves are enabled to appear in places where they are not shaded by chlorophyll. In this respect, and also in the behaviour shown by some algae where phytochrome absorption leads to movement or orientation of the chloroplast within the cell, phytochrome and chlorophyll form a cooperating pair of pigments. (Other roles of phytochrome such as in photoperiodicity and in seed germination are probably independent of chlorophyll.)

Phytochrome has been implicated in a surprising behaviour of chloroplasts in the filamentous alga *Mougeotia*. Each cell contains a single, rectangular blade-like chloroplast which turns inside the cell so as to present the optimum profile for light absorption. The responses to red and far-red light implicate phytochrome, which must be oriented and immobile with respect to the cell wall, consistent with the relationship of phytochrome to the cell membrane outlined above (see Mayer, 1971).

Chloroplast movement in relation to illumination is indeed quite common; many cells show protoplasmic streaming, and chloroplasts are observed to be captured by apparently 'sticky' regions adjacent to the cell wall. Phytochrome has not been shown to be important in most cases.

REFERENCES AND FURTHER READING

Chapter 1

Further reading

Arnon, D.I. (1977) Photosynthesis 1950–75: Changing Concepts and Perspectives. In Trebst, A. and Avron, M., eds., Photosynthesis I. Photosynthetic electron transport and phosphorylation. *Encycl. Plant Physiol.* new series, Vol. 5, Springer-Verlag, Berlin, 7–56.
Hoober, J.K. (1984) *Chloroplasts.* (Cellular Organelles) Plenum, New York, Ch. 1.
Rabinowitch, E.I. (1945) *Photosynthesis and Related Processes.* Interscience, New York, Ch. 2.
Reid, R.A. and Leech, R.M. (1980) *Biochemistry and Structure of Cell Organelles.* (Tertiary Biology Series) Blackie, Glasgow, Ch. 3.
Stoeckenius, W. and Bogolmoni, R.A. (1982) Bacteriorhodopsin and related pigments of halobacteria. *Ann. Rev. Biochem.* **52**, 587–615.

Cited references

Gabellini, N. (1988) Organisation and structure of the genes for the cytochrome b/c_1 complex in purple photosynthetic bacteria. A phylogenetic study describing the homology of the b/c_1 subunits between prokaryotes, mitochondria and chloroplasts. *J. Bioenergetics and Biomembranes* **20**, 59–83.
Riley, G.A. (1944) The carbon metabolism and photosynthetic efficiency of the earth as a whole. *American Scientist* **32**, 132–134.

Chapter 2

Further reading

*Allen, J.P., Feher, G., Yeates, T.O., Komiya, H. and Rees, D.C. (1987) Structure of the reaction centre from *Rhodobacter sphaeroides* R-26: The protein subunits. *Proc. Natl. Acad. Sci. USA* **84**, 6162–6166.
Anderson, J.M. (1987) Molecular organisation of thylakoid membranes. In Amesz, J., ed., *Photosynthesis. New Comprehensive Biochemistry* Vol. 15, Elsevier, Amsterdam, 273–297.
*Deisenhofer, J., Epp, O., Miki, K., Huber, R. and Michel, H. (1985) Structure of the protein subunits in the photosynthetic reaction centre of *Rhodopseudomonas viridis* at 3 Å resolution. *Nature* **318**, 618–624.
Hunter, C.N., van Grondelle, R. and Olsen, J.D. (1989) Photosynthetic antenna proteins: 100 ps before photochemistry starts. *Trends Biochem. Sci.* **14**, 72–76.
Pearson, B.K. and Olson, J.M. (1987) Photosynthetic bacteria. In Amesz, J., ed., *Photosynthesis. New Comprehensive Biochemistry* Vol. 15, Elsevier, Amsterdam, 21–42.
Peter, G.F., Machold, O. and Thornber, J.P. (1988) Identification and isolation of photosystem I and photosystem II pigment-proteins from higher plants. In Harwood, J.L. and Walton, T.J., eds., *Plant Membranes—Structure, Assembly and Function.* Biochemical Society, London, 17–31.
Staehelin, L.A. (1986) Chloroplast structure and supramolecular organisation of photosynthetic membranes. In Staehelin, L.A. and Arntzen, C.J., eds., *Photosynthesis III. Photosynthetic membranes and light harvesting systems. Encycl. of Plant Physiol.* new series, Vol. 19, Springer-Verlag, Berlin, 1–84.

*These articles contain coloured stereoscopic diagrams, for which a viewer will be required.

Thornber, J.P. (1986) Biochemical characterization and structure of pigment-proteins of photosynthetic organisms. In Staehelin, L.A. and Arntzen, C.J., eds., *Photosynthesis III. Photosynthetic membranes and light harvesting systems. Encycl. of Plant Physiol.* new series, Vol. 19, Springer-Verlag, Berlin, 98–142.

Zuber, H. (1986) Primary structure and function of the light-harvesting polypeptides from cyanobacteria, red algae and purple photosynthetic bacteria. In Staehelin, L.A. and Arntzen, C.J., eds., *Photosynthesis III. Photosynthetic membranes and light harvesting systems. Encycl. of Plant Physiol.* new series, Vol. 19, Springer-Verlag, Berlin, 238–251.

Zuber, H. (1986) Structure of light-harvesting antenna complexes of photosynthetic bacteria, cyanobacteria and red algae. *Trends Biochem. Sci.* 11, 414–419.

Cited references

Anderson, J.M. and Barrett, J. (1986) Light-harvesting pigment-protein complexes of algae. In Staehelin, L.A. and Arntzen, C.J., eds., *Photosynthesis III. Photosynthetic membranes and light harvesting systems. Encycl. of Plant Physiol.* new series, Vol. 19, Springer-Verlag, Berlin, 269–285.

Betti, J.A., Blankenship, R.E. Natarajan, L.A., Dickinson, L.C. and Fuller, R.C. (1982) Antenna organization and evidence for the function of a new antenna pigment species in the green photosynthetic bacterium *Chloroflexus aurantiacus. Biochim. Biophys. Acta* 680, 194–201.

Bogorad, L. (1975) Phycobilins and complementary chromatic adaptation. *Ann. Rev. Plant. Physiol.* 26, 369–401.

Branton, D., Bullivant, S., Gilula, N.B., Karnovsky, M.J., Moor, H., Mu'hlethaler, K., Northcote, D.H., Packer, L., Satir, B., Satir, P., Speth, V., Staehelin, L.A., Steere, R.L. and Weinstein, R.S. (1975) Freeze-etching nomenclature. *Science* 190, 54–56.

Chitnis, P.R. and Thornber, J.P. (1988) The major light-harvesting complex of photosystem II: aspects of its molecular and cell biology. *Photosynthesis Res.* 16, 41–63.

Deisenhofer, J., Epp, O., Miki, K., Huber, R. and Michel, H. (1984) X-ray structure analysis of a membrane protein complex. Electron density map at 3 Å resolution and a model of the chromophores of the photosynthetic reaction center from *Rhodopseudomonas viridis. J. Mol. Biol.* 180, 385–398.

Dyer, T. (1988) The molecular genetics of thylakoid proteins. In Harwood, J.L. and Walton, T.J., eds., *Plant Membranes—Structure, Assembly and Function.* Biochem. Soc., London, 139–147.

Gest, H. and Favinger, J.L. (1983) *Heliobacterium chlorum,* an anoxygenic brownish-green photosynthetic bacterium containing a 'new' form of bacteriochlorophyll. *Arch. Microbiol.* 136, 11–16.

Glazer, A.N. (1984) Phycobilisome. A macromolecular complex optimized for light energy transfer. *Biochim. Biophys. Acta* 768, 29–51.

Green, B.R. (1988) The chlorophyll-protein complexes of higher plant photosynthetic membranes *or* Just what green band is that? *Photosynthesis Res.* 15, 3–32.

Hauska, G. (1988) Phylloquinone in photosystem I: are quinones the secondary electron acceptors in all types of photosynthetic reaction centers? *Trends in Biochem. Sci.,* 13, 415–416.

Haworth, P., Watson, J.L. and Arntzen, C.J. (1983) The detection, isolation and characterization of a light-harvesting complex which is specifically associated with photosystem I. *Biochim. Biophys. Acta* 724, 151–158.

Hearst, J.E. and Sauer, K. (1984) Protein sequence homologies between portions of the L and M subunits of the reaction centres of *Rhodopseudomonas capsulata* and the 32 kDa herbicide-binding polypeptide of chloroplast thylakoid membranes and a proposed relation to quinone binding sites. *Z. Naturforsch.* 39c, 421–424.

Hurt, E.C. and Hauska, G. (1984) Purification of membrane-bound cytochromes and a photoactive P840 protein complex of the green sulfur bacterium *Chlorobium limicola f. thiosulfatophilum*. *FEBS lett.* **168**, 149–154.

Lam, E., Ortiz, W. and Malkin, R. (1984) Chlorophyll a/b proteins of photosystem I. *FEBS Lett.* **168**, 10–14.

Nanba, O. and Satoh, H. (1987) Isolation of a photosystem II reaction center consisting of D-1 and D-2 polypeptides and cytochrome b-559. *Proc. Natl. Acad. Sci. USA* **84**, 109–112.

Nelson, N. (1987) Structure and function of protein complexes in the photosynthetic membrane. In Amesz, J., ed., *Photosynthesis. New Comprehensive Biochemistry* Vol. 15, Elsevier, Amsterdam, 213–231.

Packham, N.K. and Barber, J. (1987) Structure and functional comparison of anoxygenic and oxygenic organisms. In Barber, J., ed., *The Light Reactions. Topics in Photosynthesis* Vol. 8, Elsevier, Amsterdam, 1–30.

Porter, G., Tredwell, C.J., Searle, G.F.W. and Barber, J. (1978) Photosynthetic time-resolved excitation transfer in *Porphyridium cruentum*. Pt. I: The intact alga. *Biochim. Biophys. Acta* **501**, 232–245.

Peter, G.F. and Thornber, J.P. (1988) The Antenna Pigments of photosystem II with emphasis on the major pigment-protein, LHC IIb. In Scheer and Schneider, eds., *Photosynthetic Light-Harvesting Systems* Walter de Gruyter, Berlin, 175–186.

Reed, D.W. and Clayton, R.K. (1968) Isolation of a reaction center from *Rhodopseudomonas sphaeroides. Biochem. Biophys. Res. Commun.* **30**, 471–475.

Roughan, P.G. and Slack, C.R. (1982) Cellular organisation of glycerolipid metabolism. *Ann. Rev. Plant Physiol.* **33**, 97–132.

Singer, S.J. and Nicolson, G.L. (1972) The fluid mosaic model of the structure of cell membranes. *Science* **175**, 720–731.

Smith, E.L. (1938) Solutions of chlorophyll-protein compounds (phyllochlorins) extracted from spinach. *Science* **88**, 170–171.

Sprague, S.G. and Varga, A.R. (1986) Membrane architecture of anoxygenic photosynthetic bacteria. In Staehelin, L.A. and Arntzen, C.J., eds., *Photosynthesis III. Photosynthetic membranes and light harvesting systems. Encycl. of Plant Physiol.* new series, Vol. 19, Springer-Verlag, Berlin, 603–619.

Staehelin, L.A., Golecki, J.R. and Drews, G. (1980) Submolecular organizations of chlorosomes (Chlorobium vesicles) and of their membrane attachment sites in *Chlorobium limicola. Biochim. Biophys. Acta* **589**, 30–45.

Thornber, J.P., Cogdell, R.J., Pierson, B.K. and Seftor, R.E.B. (1983) Pigment-protein complexes of purple photosynthetic bacteria: an overview, *J. Cell Biochem.* **23**, 159–169.

Witt, I., Witt, H.T., Di Fiore, D., Rögner, M., Hinrichs, W., Saenger, W., Granzin, J., Betzel, Ch. and Dauter, Z. (1988) X-ray characterization of single crystals of the reaction center I of water splitting photosynthesis. *Ber. Bunsenges. Phys. Chem.* **92**, 1503–1506.

Zuber, H., Brunisholz, R. and Sidler, W. (1987) Structure and function of light-harvesting pigment-protein complexes. In Amesz, J., ed., *Photosynthesis. New Comprehensive Biochemistry* Vol. 15, Elsevier, Amsterdam, 233–271.

Chapter 3

Further reading

Barber, J. (1987) Photochemical reaction centres: a common link. *Trends Biochem. Sci.* **12**, 321–326.

Hader, D-P. and Tevini, M. (1987) Excited states and their relaxation *and* Photochemical reactions and energy transfer. *General Photobiology* Pergamon, Oxford, Ch. 5 and 6.

Knaff, D.B. (1988) The photosystem I reaction centre. *Trends Biochem. Sci.* **13**, 460–461.

Parson, W.W. (1987) The bacterial reaction center. In Amesz, J., ed., *Photosynthesis. New Comprehensive Biochemistry* Vol. 15, Elsevier, Amsterdam, 43–61.

Youvan, D.C. and Marrs, B.L. (1987) Molecular mechanisms of photosynthesis. *Scientific American* **256** (6), June, 42–48.

Cited references

Deisenhofer, J., Epp, O., Miki, K., Huber, R. and Michel, H. (1985) Structure of the protein subunits in the photosynthetic reaction centre of *Rhodopseudomonas viridis* at 3 Å resolution. *Nature* **318**, 618–624.

Dörnemann, D. and Senger, H. (1985) Optical properties and structure of chlorophyll RC I. In Blauer, G. and Sund, H., eds., *Optical properties and structure of tetrapyrroles* de Gruyter, Berlin, 488–505.

van Dorssen, R.J., Plijter, J.J., den Ouden, A., Amesz, J. and van Gorkom, H.J. (1987) Pigment arrangement in photosystem II. In Biggins, J. ed., *Progress in Photosynthesis Research, Proc. Int. Congr. Photosynth. Res., VIIth 1986* Martinus Nijhoff, Dordrecht, Vol. 1, 439–442.

Giorgi, L.B., Gore, B.L., Klug, D.R., Ide, J.P., Barber, J. and Porter, G. (1987) Picosecond transient absorption spectroscopy of photosystem I reaction centres from higher plants. In Biggins, J., ed., *Progress in Photosynthesis Research, Proc. Int. Congr. Photosynth. Res., VIIth 1986* Martinus Nijhoff, Dordrecht, Vol. 1, 257–260.

Mansfield, R.W. and Evans, M.C.W. (1985) Optical difference spectrum of the electron acceptor A_0 in photosystem I. *FEBS Lett.* **190**, 237–241.

Nuijs, A.M., van Dorssen, R.J., Duysens, L.N.M. and Amesz, J. (1985) Excited states and primary photochemical reactions in the photosynthetic bacterium *Heliobacterium chlorum*. *Proc. Natl. Acad. Sci.*, USA. **82**, 6965–6968.

Nuijs, A.M., Shuvalov, V.A., Smith, H.W.J., van Gorkom, H.J. and Duysens, L.N.M. (1987) Excited states and primary photochemical reactions in photosystem I. In Biggins, J. ed., *Progress in Photosynthesis Research. Proc. Int. Congr. Photosynth. Res., VIIth 1986* Martinus Nijhoff, Dordrecht, Vol. 1, 229–232.

Parson, W.W. (1987) see Chapter 4.

Strain, H.H. and Svec, W.A. (1966) Extraction, separation, estimation and isolation of the chlorophylls. In Vernon, L.P. and Seely, G.R., eds, *The Chlorophylls* Academic Press, New York, 22–66.

Thornber, J.P. (1970) Photochemical reactions of purple bacteria as revealed by studies of three spectrally different carotenobacteriochlorophyll-protein complexes isolated from *Chromatium*, Strain D. *Biochemistry* **9**, 2688–2698.

Watanabe, T., Kobayashi, M., Nakazato, M., Ikegami, I. and Hiyama, T. (1987) Chlorophyll a' in photosynthetic apparatus: reinvestigation. In Biggins, J., ed., *Progress in Photosynthesis Research, Proc. Int. Congr. Photosynth. Res., VIIth 1986* Martinus Nijhoff, Dordrecht, Vol. 1, 303–306.

Wittmershaus, B.P. (1987) Measurements and kinetic modelling of picosecond time-resolved fluorescence from photosystem I and chloroplasts. In Biggins, J., ed., *Progress in Photosynthesis Research, Proc. Int. Congr. Photosynth. Res., VIIth 1986* Martinus Nijhoff, Dordrecht, Vol. 1, 75–82.

Chapter 4

Further reading

Andréasson, L.E. and Vänngård, T. (1988) Electron transport in photosystems I and II. *Ann. Rev. Plant Physiol.* **39**, 379–41.

Mathis, P. and Rutherford, A.W. (1987) The primary reactions of photosystems I and II of algae and higher plants. In Amesz, J., ed., *Photosynthesis. New Comprehensive Biochemistry* Vol. 15 Elsevier, Amsterdam, 63–96.

Parson, W.W. (1987) The bacterial reaction center. In Amesz, J., ed., *Photosynthesis. New Comprehensive Biochemistry* Vol. 15 Elsevier, Amsterdam, 43–61.

Cited references

Barber, J. (1987) Photosynthetic reaction centres: a common link. *Trends Biochem. Sci.* **12**, 321–326.

Brudvig, G.W. and de Paula, J.C. (1987) on the mechanism of photosynthetic water oxidation. In Biggins, J., ed., *Progress in Photosynthesis Research, Proc. Int. Congr. Photosynth. Res., VIIth 1986* Martinus Nijhoff, Dordrecht, Vol. 1, 491–498.

Davenport, H.E. (1972) Some observations on cytochrome *f*. In Forti, G., Avron, M. and Melandri, A., eds., *Proc Int. Congr. Photosynth. Res., IInd 1971* 1593–1601.

Davenport, H.E. and Hill, R. (1952) The preparation and some properties of of cytochrome *f*. *Proc. Roy. Soc. ser B* **139**, 327–345.

Dutton, P.L. (1986) Energy Transduction in anoxygenic photosynthesis. In Staehelin, L.A. and Arntzen, C.J., eds., *Photosynthesis III. Photosynthetic membranes and light harvesting systems. Encycl. of Plant Physiol.* New series, Vol. 19, Springer-Verlag, Berlin, 197–237.

Guss, J.M. and Freeman, H.C. (1983) Structure of oxidised poplar plastocyanin at 1.6 Å resolution. *J. Mol. Biol.* **169**, 521–563.

Holten, D., Windsor, M.W., Parson, W.W. and Thornber, J.P. (1978) *Biochim. Biophys. Acta* **501**, 112–116.

Joliot, P., Barbieri, G. and Chabaud, R. (1969) Un nouveau modè le des centres photochemiques du système II. *Photochem. Photobiol.* **10**, 309–329.

Kok, B., Forbush B. and McGloin, M. (1970) Cooperation of charges in photosynthetic oxygen evolution—I. A linear four-step mechanism. *Photochem. Photobiol.* **11**, 457–475.

Lavorel, J. and Etienne, A-L. (1977) In vivo chlorophyll fluorescence. In Barber, J., ed., Primary Processes in Photosynthesis, *Topics in Photosynthesis* Vol. 2, Elsevier/North Holland Biomedical Press, 203–268.

Mansfield, R.W., Nugent, J.H.A. and Evans, M.C.W. (1987) Investigation of the chemical nature of the electron Acceptor A_1 in photosystem I of higher plants. In Biggins, J., ed., *Progress in Photosynthesis Research, Proc. Int. Congr. Photosynth. Res., VIIth 1986* Martinus Nijhoff, Dordrecht, Vol. 1, 241–247.

Nuijs, A.M., Shuvalov, V.A., Smit, H.W.J., van Gorkom, H.J. and Duysens, L.N.M. (1987) Excited states and primary photochemical reactions in photosystem I. In Biggins, J., ed., *Progress in Photosynthesis Research, Proc. Int. Congr. Photosynth, Res., VIIth 1986* Martinus Nijhoff, Dordrecht, Vol. 1, 229–232.

Parson, W.W. (1969) The reactors between primary and secondary electron acceptors in bacterial photosynthesis. *Biochim. Biophys. Acta* **189**, 384–396.

Sétif, P. and Mathis, P. (1980) The oxidation-reduction potential of P700 in chloroplast lamellae and sub-chloroplast particles. *Arch. Biochem. Biophys.* **204**, 477–485.

Trebst, A. (1986) The topology of the plastoquinone and herbicide binding peptides of photosystem II in the thylakoid membrane. *Z. Naturforsch.* **41c**, 240–245.

Trebst, A. (1988) Herbicide action of photosynthetic membranes. In Harwood, J.L. and Walton, T.J., eds., *Plant Membranes—Structure, Assembly and Function* Biochem. Soc., London, 201–208.

Chapter 5

Further reading

Hauska, G. (1986) Composition and structure of cytochrome bc_1 and b_6f complexes. In Staehelin, L.A. and Arntzen, C.J., eds., *Photosynthesis III. Photosynthetic membranes and light harvesting systems. Encycl. of Plant Physiol.* new series, Vol. 19, Springer-Verlag, Berlin, 496–507.

O'Keefe, D.P. (1988) Structure and function of the chloroplast cytochrome *bf* complex. *Photosynthesis Res.* **17**, 189–216.

Cited references

Anderson, J.M. (1987) Molecular organisation of thylakoid membranes. In Amesz, J., ed., *Photosynthesis. New Comprehensive Biochemistry* Vol. 15 Elsevier Amsterdam, 273–297.

Barber, J. (1984) Further evidence for the ancestry of cytochrome $b-c$ complexes. *Trends Biochem. Sci.* **9**, 209–211.
Mitchell, P. (1976) Possible molecular mechanisms of the protonmotive function of cytochrome systems. *J. Theor. Biol.* **62**, 327–367.
Pschorn, R., Rühle, W. and Wild, A. (1988) Structure and function of ferredoxin-$NADP^+$-oxidoreductase. *Photosynthesis Res.* **17**, 217–229.

Chapter 6

Further reading

Nicholls, D.G. (1982) *Bioenergetics. An introduction to the chemiosmotic theory* Academic Press, London, Chapters 6 and 7.
Ort, D.R. and Good, N.E. (1988) Textbooks ignore photosystem II-dependent ATP formation: is the Z-scheme to blame? *Trends Biochem. Sci.* **13**, 467–469.

Cited references

Arnon, D.I., Allen, M.B. and Whatley, F.R. (1954) Photosynthesis by isolated chloroplasts. *Nature* **174**, 394–396.
Chance, B. and Williams, G.R. (1955) Respiratory enzymes in oxidative phosphorylation. III, The steady state. *J. Biol. Chem.* **217**, 409–427.
Godde, D. (1982) Evidence for a membrane bound NADH-plastoquinone-oxidoreductase in *Chlamydomonas reinhardtii* CW-15. *Arch. Microbiol.* **131**, 197–202.
Hind, G. and Jagendorf, A.T. (1963) *Proc. Natl. Acad. Sci. USA.* **49**, 715–722.
Kouyama, T., Nasuda-Kouyama, A., Ikegami, A., Matthew, M.K. and Stoeckenius, W. (1988) Bacteriorhodopsin photoreaction: identification of a long-lived intermediate N(P, R_{350}) at high pH and its M-like photoproduct. *Biochemistry* **27**, 5855–5863.
Mathies, R.A., Brito Cruz, C.H., Pollard, W.T. and Shank, C.V. (1988). Direct observation of the femtosecond excited-state *cis–trans* isomerization in bacteriorhodopsin. *Science* **240**, 777–779.
Mitchell, P. (1961) Coupling of phosphorylation to electron and hydrogen transfer by a chemiosmotic type of mechanism. *Nature* **191**, 423–427.
Stoeckenius, W. (1985) The rhodopsin-like pigments of halobacteria: light-energy and signal transducers in an archaebacterium. *Trends Biochem. Sci.* **10**, 483–486.
Stryer, L. (1988) *Biochemistry*, 3rd ed., WH Freeman & Co., New York.
West, K.R. and Wiskich, J.T. (1968) Photosynthetic control by isolated pea chloroplasts. *Biochem. J.* **109**, 527–532.
Wraight, C.A., Cogdell, R.J. and Chance, B. (1978) in Clayton, R.K. and Sistrom W.R., eds., *The Photosynthetic Bacteria* Plenum Press, New York, 471–511.

Chapter 7

Further reading

Edwards, G.E. and Walker, D.A. (183) *C3, C4: mechanisms, and cellular and environmental regulation, of photosynthesis* Blackwell, Oxford, part B, 107–495.
Halliwell, B. (1984) *Chloroplast Metabolism. The structure and function of chloroplasts in green leaf cells* Oxford University Press, Chapters 4, 6 and 7.
Macdonald, F.D. and Buchanan, B.B. (1987) Carbon dioxide assimilation. In Amesz, J., ed., *Photosynthesis. New Comprehensive Biochemistry* Vol. 15, Elsevier, Amsterdam, 175–197.
Ohyama, K., Kohchi, T., Sano, T. and Yamada, Y. (1988) Newly identified groups of genes in chloroplasts. *Trends Biochem. Sci.* **13**, 19–22.

152   PHOTOSYNTHESIS

Cited references

Bennet, J. (1981) Biosynthesis of the light-harvesting chlorophyll a/b protein. *Europ. J. Biochem.* **118**, 61–70.

Berry, J.A., Lorimer, G.H., Pierce, J., Seemann, J.R., Meek, J. and Freas, S. (1987) Isolation, identification and synthesis of 2-carboxyarabinitol-1-phosphate, a diurnal regulator of ribulose-bisphosphate carboxylase activity. *Proc. Natl. Acad. Sci., USA* **84**, 734–738.

Black, C.C. and Mollenhauer, H.H. (1971) Structure and distribution of chloroplasts and other organelles in leaves with various rates of photosynthesis. *Plant Physiol.* **47**, 15–23.

Bowes, G., Ogren, W.L. and Hagemen, R.H. (1971) Phosphoglycollate production catalysed by ribulose diphosphate carboxylase. *Biochem. Biophys. Res. Commun.* **45**, 716–722.

Crouse, E. J., Schmitt, J.M. and Bohnert, H-J. (1985) Chloroplast and cyanobacterial genomes, genes and RNAs: a compilation. *Plant Mol. Biol. Reporter* **3**, 43–89.

Douce, R. and Joyard, J. (1979) Structure and function of the plastid envelope. *Adv. Bot. Res.* **7**, 1–116.

Ellis, R.J. (1977) Protein synthesis by isolated chloroplasts. *Biochim. Biophys. Acta* **463**, 185–215.

Harwood, J. (1988) The site of action of some selective graminaceous herbicides is identified as acetyl-CoA carboxylase. *Trends Biochem. Sci.* 330–331.

Husic, D.W., Husic, H.D. and Tolbert, N.E. (1987) The oxidative Carbon cycle or C_2 cycle. *Critical Revs. Plt. Sci.* **5**, 45–100.

Kelly, G.J. (1984) The capture of carbon by aquatic plants. *Trends Biochem. Sci.* **9**, 255–256.

Mayer, F. (1971) Light-induced chloroplast contraction and movement. In Gibbs, M., ed., *Structure and Function of Chloroplasts* Springer-Verlag, Berlin, 35–49.

Mullet, J.E. (1988) Chloroplast development and gene expression. *Ann. Rev. Plant Physiol.* **39**, 475–502.

Schuster, G., Schochat, S., Adir, N., Even, D., Ish-Shalom, D., Grimm, B., Kloppstech, K. and Ohad, I. (1988) The synergistic effect of light and heat stress on the inactivation of photosystem II. In Harwood, J.L. and Walton, T.J., eds., *Plant Membranes—Structure, Assembly and Function* Biochemical Society, London, 133–138.

Ohyama, K., Kohchi, T., Fukuzawa, H., Sano, T., Umesono, K. and Ozeki, H. (1988) Gene organisation and newly identified groups of genes of the chloropast genome from a liverwort, *Marchantia polymorpha*. *Photosynthesis Res.* **16**, 7–22.

Salvucci, M.E., Portis, A.R. and Ogren, W.L. (1986) A soluble chloroplast protein catalyses ribulosebisphosphate carboxylase/oxygenase activation in vivo. *Photosynthesis Res.* **7**, 193–201.

Shinozaki, K., Hayashida, N. and Sugiura, M. (1988) Nicotiana chloroplast genes for components of the photosynthetic apparatus. *Photosynthesis Res.* **18**, 7–31.

Woodrow, I.E. and Berry, J.A. (1988) Enzymic regulation of photosynthetic CO_2 fixation in C3 plants. *Ann. Rev. Plant Physiol.* **39**, 533–594.

Zelitch, I. (1979) Photorespiration: studies with whole tissues. In Gibbs, M. and Latzko, eds., *Photosynthesis II. Photosynthetic Carbon Metabolism. Encycl. Plant Physiol.* new series, Vol. 6, Springer-Verlag, Berlin, 174–180.

Index

Hales, S. 3
herbicides 70−1, 134
heterocysts 113
heterotrophy 6
Hill, R. 3, 4, 40
Hill reaction 3, 108
Hind, G. 96
histidine, in proteins 35, 57, 65
histones, and nuclear DNA 142
history; time-charts 3, 40, 96
Hordeum vulgare, barley 10
HQNO, NQNO, inhibitors of *bc* complexes 87
hydrazine, donor to PSII 77
hydrogen, H_2
  redox potential 62
  as electron donor 88
  production 7, 111−13
hydrogen electrode, standard 61−4
hydrogen ions, H^+ (*see also* protons)
  and ATP production 8, 85, 88*ff*
  gradient (*see also* PMF) 95−7, 97−107
  localised channels 106
  and reverse electron transport 88
hydrogen peroxide 114
hydrogen sulphide 5, 62, 81, 88, 94, 110
hydroxylamine, alternative donor to PSII 77

inductive resonance 51
Ingen-Housz, J. 3, 4, 13
Ioxynil, inhibitor of PSII 71
iron, in reaction centres 59, 62, 68, 74
iron-sulphur centres
  in Chlorobiaceae 75
  see also ferredoxin
  in *Heliobacterium* 75
  in inorganic reductions 111−11
  in PSI; A, B and X 44, 58, 62, 72−5, 81
  see also Rieske centre

Jagendorf, A.T. 96

Kalckar, H. 96
Kautsky effect 69
Kessler, E. 76
α-ketoglutarate, *see* 2-oxoglutarate
Kok, B. 40, 55, 133
Kranz anatomy 128−9

L-subunit of purple bacteria RC 42, 57−9, 67, 76, 80
Lavoisier, A. 3, 4
leader sequence, of protein precursor 142
Lewis, C.M. 40
light harvesting complex
  see antenna
  see LHC

LHCI 36−9, 43
LHCII 35−9, 43, 52−3, 55
linker proteins 31, 34, 38, 42
Lipmann, F. 96
lipids 16−17, 134
lithotrophs 5−6
Lohmann, K. 96
lutein 38−39, 48
lysine, in proteins and retinal 98
Lyubimova, M.N. 96

M-subunit, of purple bacteria RC 41, 57−9, 67, 76, 80
magnesium 25, 122
maize, C4 in 129*ff*
malate, metabolism 5−6, 116−17, 129−32
  transporter 134−6
malic enzymes (NAD- and NADP-ME) 130-1
manganese 42, 76−80, 81
Marchantia polymorpha 10, 142
Martin, A.J.P. 118
mass spectrograph, for $^{18}O_2$ 123
Mastigocladus laminosus 10, 33
membrane 13*ff*, 16−17, 19−24
membrane potential, and PMF 99−100
membrane-spanning proteins 14, 142
  see bacteriorhodopsin
  cytochrome *bc*, b_6f complexes 84−5
  purple bacteria RC 57
  purple bacteria antenna 35
  in PSI 43, 74
  see also PSII
menaquinone (Vitamin K_2) 19, 41, 67, 85, 95
mesophyll 139
  of C4 plants 128*ff*
^{35}S-methionine, and new protein 144
Meyer, A. 3
Meyer, R. 3
Mitchell, P. 95, 96, 100
mitochondrion 11, 12, 15, 19, 20−1, 133
  and ATP production 7, 8, 97, 100, 118
  control by chloroplast 118
  cytochromes of 75, 90
  in photorespiration 124−5, 133
  PMF value in 103
  porins of 134
von Mohl, H. 3
molecular cell biology 140−4
molybdenum, in N-metabolism 112−13
monogalactosyldiacylglycerol (MGDG) 16−17
Mössbauer spectra, technique 73
Mougeotia 10, 145